ON SITE GEOARCHAEOLOGY ON A NEOLITHIC TELL SITE IN GREECE: ARCHAEOLOGICAL SEDIMENTS, MICROARTIFACTS AND SOFTWARE DEVELOPMENT

On Site Geoarchaeology on a Neolithic Tell Site in Greece: Archaeological Sediments, Microartifacts and Software Development

Dimitris Kontogiorgos
Editor

Nova Science Publishers, Inc.
New York

Copyright © 2009 by Nova Science Publishers, Inc.

All rights reserved. No part of this book may be reproduced, stored in a retrieval system or transmitted in any form or by any means: electronic, electrostatic, magnetic, tape, mechanical photocopying, recording or otherwise without the written permission of the Publisher.

For permission to use material from this book please contact us:
Telephone 631-231-7269; Fax 631-231-8175
Web Site: http://www.novapublishers.com

NOTICE TO THE READER

The Publisher has taken reasonable care in the preparation of this book, but makes no expressed or implied warranty of any kind and assumes no responsibility for any errors or omissions. No liability is assumed for incidental or consequential damages in connection with or arising out of information contained in this book. The Publisher shall not be liable for any special, consequential, or exemplary damages resulting, in whole or in part, from the readers' use of, or reliance upon, this material.

Independent verification should be sought for any data, advice or recommendations contained in this book. In addition, no responsibility is assumed by the publisher for any injury and/or damage to persons or property arising from any methods, products, instructions, ideas or otherwise contained in this publication.

This publication is designed to provide accurate and authoritative information with regard to the subject matter covered herein. It is sold with the clear understanding that the Publisher is not engaged in rendering legal or any other professional services. If legal or any other expert assistance is required, the services of a competent person should be sought. FROM A DECLARATION OF PARTICIPANTS JOINTLY ADOPTED BY A COMMITTEE OF THE AMERICAN BAR ASSOCIATION AND A COMMITTEE OF PUBLISHERS.

LIBRARY OF CONGRESS CATALOGING-IN-PUBLICATION DATA

Kontogiorgos, Dimitris.
  On site geoarchaeology on a Neolithic tell site in Greece : archaeological sediments, microartifacts, and software development / Dimitris Kontogiorgos.
      p. cm.
   Includes index.
   ISBN 978-1-60741-366-0 (hardcover)
  1. Paliambela Kolindros Site (Greece) 2. Neolithic period--Greece--Pieria Region. 3. Excavations (Archaeology)--Greece--Pieria Region. 4. Archaeological geology--Greece--Pieria Region. 5. Archaeological geology--Computer programs. 6. Sediments (Geology)--Analysis. 7. Particle size determination. 8. Soil science in archaeology--Greece--Pieria Region. 9. Antiquities, Prehistoric--Greece--Pieria Region. 10. Pieria Region (Greece)--Antiquities. I. Title.
  GN776.22.G8K66 2009
  938'.1--dc22
                              2009017368

Published by Nova Science Publishers, Inc. ✦ *New York*

CONTENTS

Acknowledgements   vii

Chapter 1   The Rise of a Neolithic Tell: The Evidence from Particle Size Analysis on Archaeological Sediments   1
Dimitris Kontogiorgos

Chapter 2   Tell the Story of a Ditch. Additional Sedimentary and Microartifactual Evidence for the Use of Space on a Neolithic Tell   19
Dimitris Kontogiorgos

Chapter 3   Test again for the Optimum! Confirming the Use of Genetic Algorithm Minimization on Microartifact Weight Estimation   27
Dimitris Kontogiorgos and Alexandros Leontitsis

Chapter 4   Microartelyzer: A Software for Microartifacts' Quantification   47
Alexandros Leontitsis and Dimitris Kontogiorgos

Chapter 5   From Tells to Extended Settlements: On-site Geoarchaeology and Cultural Formation Processes at the Extended Neolithic Settlement at Korinos (Northern Greece)   57
Manthos Besios, Fotini Adaktylou and Dimitris Kontogiorgos

Chapter 6	From Neolithic to Hellenistic. A Geoarchaeological Approach to the Burial of a Hellenistic Theatre: The Evidence from Particle Size Analysis and Microartifacts *Dimitris Kontogiorgos and Kaliopi Preka-Alexandri*	**71**
Chapter 7	A Comment on the Vertical Movement of Microartifacts in an Anthropogenic and in a Non Anthropogenic Environment and the Effects of Formation Processes *Dimitris Kontogiorgos*	**81**
Chapter 8	Concluding Remarks *Dimitris Kontogiorgos*	**85**
Index		**89**

ACKNOWLEDGEMENTS

Dimitris Kontogiorgos would like to express his deep appreciation and gratitude to Prof. Paul Halstead (University of Sheffield-U.K.) for his valuable academic advice, support and encouragement. D.K would like also to thank Professor Kostas Kotsakis (Aristotle University of Thessaloniki-Greece) for his encouragement to study the geoarchaeology of the site. D.K would like also to express his appreciation to Dr. Sherry Fox (Wiener Laboratory at the American School of Classical studies at Athens) for her support. This work would not be possible without generous funding from the Wiener Laboratory at the American School of Classical Studies at Athens.

In: On Site Geoarchaeology on a Neolithic Tell Site ... ISBN 978-1-60741-366-0
Editor: D. Kontogiorgos © 2009 Nova Science Publishers, Inc.

Chapter 1

THE RISE OF A NEOLITHIC TELL: THE EVIDENCE FROM PARTICLE SIZE ANALYSIS ON ARCHAEOLOGICAL SEDIMENTS

Dimitris Kontogiorgos
Department of Archaeology, University of Sheffield, Northgate House, Sheffield, U.K.

1.1. INTRODUCTION

In Southeast Europe, the majority of tell sites are in the areas adjacent to or geographically close to the Near East, such as Bulgaria, Greece, Southern Romania, and in the Central Balkans, in Serbia as far north as the Danube river (e.g., Lloyd, 1963). In terms of architecture, tell settlements seem to offer an uninterrupted sequence of living horizons, and good preservation of houses, and consequently more attention archaeologically (c.f., Davidson, 1976).

Three important elements in tell formation have been identified by Sherratt (1983) - the intensive use of mud for building, a high degree of locational stability, and the concentration of houses in a coherent unit. Rosen's discussion on the development of Israeli tells, clearly demonstrates that tell formation is a complex process under normal conditions of occupation, expansion, and rebuilding (Rosen, 1986). Chapman (1989) has suggested another type of site, that may resemble a tell, and is formed due to the constraints posed in a settlement's expansion by its location (i.e., on a hill) which enforces concentration of houses.

In addition, Chapman (op.cit) talks about the significance of earth/mud used in house construction for the formation of tells. Stevanovic and Tringham (1997) have pointed out that the strategy of house replacement is an important factor in tell formation, and acknowledge that the use of earth/mud for construction intensified during the Neolithic period. Halstead (1999) and Kotsakis (1999) sustain the idea of successive rebuilding of houses inside tells, and consider, as a basic precondition for tell formation, the use of mud-brick.

In general, studies on tell formation argue that the primary component of tell sediments is eroded mud-brick from structures, with the addition of eroded organic and cultural debris, collapse rubble, sheet-wash sediments, biogenic or geo-chemically altered sediments, and some natural alluvial and aeolian sediments, particularly after abandonment (e.g., Davidson, 1973; 1976; Rosen, 1986). Textural data, on a recent geoarchaeological study conducted on the Neolithic tell site at Paliambela in Northern Greece (Figure 1.1), indicated that the tell material was produced by introducing alluvial material and an additional contribution of coarser sands and finer gravel fractions (i.e., small granule to small pebbles) (Kontogiorgos, 2008).

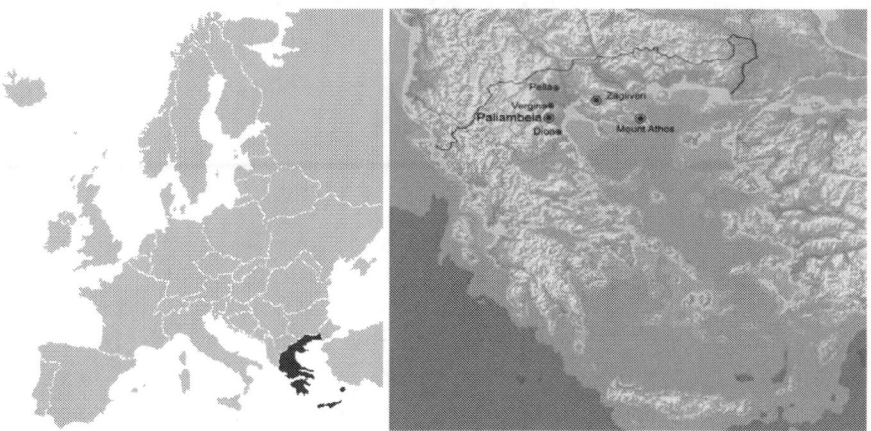

Source: 'Paliambela excavation' archive.

Figure 1.1. a) Map of Europe, showing the location of Greece, b) Map of Greece, showing the location of Paliambela.

In this study, particle size distributions of sediments from occupation deposits and building materials (i.e., adobe) of the Neolithic tell site at Paliambela as well as locally available sediments[1] (i.e., Neogene sandy loams and Holocene alluvial deposits) will be compared, in an attempt to further explore the origin of tell material on the basis of textural parameters. The first part of this study describes the laboratory procedures, while the second part presents and interprets the particle size data. Finally, the third part offers some concluding remarks.

1.2. LABORATORY PROCEDURES

1.2a. Particle-size Analysis (P.S.A) (Hydrometer and Sieve Analysis)

A total of one hundred eight (108) sediment samples derived from three Cores, drawn from the extensive coring conducted at the Neolithic Tell and eight (8) samples of adobe material, macroscopically judged to be different in textural characteristics[2], were analyzed for their particle size composition. Hydrometer analysis was used for silt and clay, and sieving for the coarse particles (gravel and sand). Hydrometer analysis, often used for silt and clay, is based on the assumption that particles of different size settle in a column of water at a rate directly proportional to their diameter. Fine particles will take longer to settle than coarse particles (Folk, 1980). Only the material, which passed through a 2mm sieve, was used for hydrometer analysis.

The remaining fine-grained samples were split to a weight of about 40gr and hydrated with a dispersant (50ml of sodium hexametaphosphate [NaPo3] per sample). The prepared sample was placed into a 1000 ml sedimentation cylinder and suspended in distilled water (1000 ml). It was shaken for one minute and allowed to settle undisturbed. ASTM (1961) procedure was used: a hydrometer is inserted to measure the density of the suspension at specific points in time (corresponding to certain particle sizes) (Kaddah, 1974: 104-105). Readings were taken after 1, 3, 15, and 45 minutes. Additional readings were taken after 2, 5, and 24 hours. All determinations were done in a constant temperature room at 20° ± 0.5° C. This observation (i.e., temperature) is of critical importance since the data are used to draw a summation percentage curve from which the percentage of different size particles is determined.

The procedure is based upon Stokes Law (see Galehouse, 1971 for some theoretical considerations on the application of Stoke's Law) and for this reason it

---

[1] The data on natural sediments were provided by Dr. Krahtopoulou, N.
[2] Kaltsogianni, pers.comm. The adobe samples were provided by Kaltsogianni S.

is necessary to know the temperature at the time of measurement since this affects the viscosity of the solution. After the 24 hour reading, the contents of the settling tube were poured through a 63μ sieve, washed and dried in a drying oven. Once dried, they were analyzed for the sand fraction, by sieving.

The most common means of determining the size distribution of sediments coarser than silt is sieving. A series of wire mesh screens are used to break a sample into different size fractions. The weight of each fraction is recorded and then analyzed statistically. The procedures involved in particle-size analysis by sieving are relatively straightforward, and the results have been shown to be highly reproducible (Rogers, 1965).

For the purpose of this study, the gravel fraction was sieved independently of the sand by using a stack of sieves from -5ϕ to -1ϕ (i.e., ϕ or phi scale is the logarithmic transformation of millimeters). The subdivision of gravel material will be used in order to address questions regarding the potential sources of coarser materials in the examined deposits. For the sand fraction a stack of sieves from 0ϕ to 4ϕ (i.e., the break between sieve and hydrometer analysis, that is, at 63μ) was used. After sieving, the mass in each sieve was weighed.

1.2b. P.S.A Statistics

Work on describing the particle size distribution itself started with the representation of observed data by Gram-Charlier series and Pearson curves (Otto, 1939; Tanner, 1958). A formulation of distributional models came with the use of the log-normal distribution (Krumbein, 1938; Krumbein and Pettijohn, 1938). This was initially a pragmatic choice, though some justification is available through the law of breakage (Kolmogorov, 1941) and was supported by some empirical evidence. These models fit many data sets moderately well but systematic discrepancies were revealed later with the availability of higher quality data and improved graphical presentations (Barndorff-Nielsen (1977), Bagnold and Barndorff-Nielsen (1980), following earlier empirical work by Bagnold (1937, 1941)). This led Bagnold and Barndorff-Nielsen (1980) to propose the log-hyperbolic distribution as a suitable model for particle size data. However, computational difficulties in fitting this distribution make the use of a simplified version, the log-skew-Laplace distribution, desirable (Olbricht, 1982; Fieller et al., 1984). Fieller et al. (1992) in their analyses of samples from Oronsay illustrate that the log-skew-Laplace model provides an adequate fit to the data which is better than that of the log-normal model and is not appreciably different from that of the log-hyperbolic model.

Table 1.1. Cumulative percentages: Adobe (Samples 1-8) and natural deposits (Avrg.)

	sample 1	sample 3	sample 6	sample 8	Sample 7	sample 5	sample 4	sample 2	Neogene sandy loam no. 1	Neogene sandy loam no.2	Holocene alluvial deposits (pre-MN)	Holocene alluvial deposits (pre-MN)	Holocene alluvial deposits (pre-MN)	Holocene alluvial deposits (Meso-lithic)	Holocene alluvial deposits (Meso-lithic)
-5.00	0.0	0.0	0.0	0.0	0.0	0.0	0.0	0.0	0.0	0.0	0.0	0.0	0.0	0.0	0.0
-4.00	0.0	0.0	0.0	0.0	0.0	0.0	0.0	0.0	0.0	0.0	0.0	0.0	0.0	0.0	0.0
-3.00	4.4	7.3	5.0	0.0	4.6	2.2	0.7	0.0	0.2	0.2	0.0	0.0	0.0	0.0	0.0
-2.00	8.8	7.6	5.0	1.1	4.8	3.3	0.9	0.0	2.1	2.0	1.5	0.7	0.4	0.4	2.0
-1.00	10.3	7.6	5.0	1.4	4.9	3.6	0.9	0.0	4.9	4.7	2.8	1.4	0.7	1.0	3.6
-0.50	20.1	27.4	15.0	15.3	12.5	18.3	28.7	17.8	5.1	5.8	3.2	1.7	0.8	1.1	3.9
0.00	30.5	41.8	21.6	29.2	21.1	30.9	39.4	30.7	6.8	9.4	4.3	2.5	1.4	1.6	5.1
0.50	39.9	55.6	26.8	41.2	28.2	38.5	52.8	42.3	10.3	16.4	6.1	3.7	2.3	2.5	7.1
1.00	49.1	65.9	32.5	51.9	37.1	46.1	63.3	53.2	22.0	32.4	8.6	5.4	3.4	4.4	10.8
1.50	59.4	77.1	38.2	61.6	45.6	54.4	71.7	66.6	51.6	58.9	11.6	7.4	4.5	7.9	15.9
2.00	68.1	90.0	43.9	69.9	54.9	61.3	80.1	78.7	66.1	71.0	15.8	10.1	5.9	14.5	22.7
2.50	74.7	96.1	48.2	75.5	66.4	69.1	85.9	85.2	71.5	75.1	20.3	13.0	7.7	21.6	28.3
3.00	83.6	97.6	57.7	84.7	78.5	78.2	92.3	91.1	77.7	79.2	25.1	16.5	11.0	28.8	33.5
3.50	84.1	97.8	58.2	84.7	78.8	78.5	92.6	91.3	77.9	79.4	28.7	19.4	14.3	33.7	37.4
4.00	86.5	98.5	62.5	88.0	80.9	81.0	94.1	92.6	79.3	79.6	32.1	22.7	18.7	38.1	41.2
5.06	93.3	90.2	73.9	90.8	82.1	86.1	91.3	86.4	95.2	98.8	55.8	50.6	49.4	65.0	68.1
6.20	94.4	95.1	83.4	97.7	92.8	91.1	96.3	93.8	95.2	98.8	64.6	58.7	57.6	71.2	74.9
6.99	98.9	98.8	88.1	100.0	96.4	96.2	98.8	97.5	96.4	99.9	69.4	64.6	64.6	75.8	78.5
7.72	100.0	100.0	95.3	100.0	97.6	98.7	100.0	100.0	98.8	99.9	76.4	72.7	71.6	79.6	82.1
8.38	100.0	100.0	100.0	100.0	100.0	100.0	100.0	100.0	98.8	99.9	80.3	76.7	75.7	83.6	85.3
9.49	100.0	100.0	100.0	100.0	100.0	100.0	100.0	100.0	100.0	99.9	85.5	81.8	81.1	87.0	88.5
15.00	100.0	100.0	100.0	100.0	100.0	100.0	100.0	100.0	100.0	100.0	100.0	100.0	100.0	100.0	100.0

Table 1.2. Cumulative percentages: Adobe and Bedrock in Core 83

sample 1	sample 3	sample 6	sample 8	sample 7	sample 5	sample 4	sample 2	Bedrock PC83	Bedrock PC 83
0.0	0.0	0.0	0.0	0.0	0.0	0.0	0.0	0.0	0.0
0.0	0.0	0.0	0.0	0.0	0.0	0.0	0.0	1.2	3.2
4.4	7.3	5.0	0.0	4.6	2.2	0.7	0.0	4.2	9.8
8.8	7.6	5.0	1.1	4.8	3.3	0.9	0.0	8.1	15.6
10.3	7.6	5.0	1.4	4.9	3.6	0.9	0.0	11.8	20.6
20.1	27.4	15.0	15.3	12.5	18.3	28.7	17.8	14.2	23.7
30.5	41.8	21.6	29.2	21.1	30.9	39.4	30.7	17.3	27.5
39.9	55.6	26.8	41.2	28.2	38.5	52.8	42.3	20.3	30.5
49.1	65.9	32.5	51.9	37.1	46.1	63.3	53.2	24.0	33.8
59.4	77.1	38.2	61.6	45.6	54.4	71.7	66.6	27.7	37.4
68.1	90.0	43.9	69.9	54.9	61.3	80.1	78.7	31.4	40.8
74.7	96.1	48.2	75.5	66.4	69.1	85.9	85.2	34.7	43.9
83.6	97.6	57.7	84.7	78.5	78.2	92.3	91.1	40.3	51.9
84.1	97.8	58.2	84.7	78.8	78.5	92.6	91.3	42.7	52.4
86.5	98.5	62.5	88.0	80.9	81.0	94.1	92.6	45.7	55.4
93.3	90.2	73.9	90.8	82.1	86.1	91.3	86.4	67.3	69.9
94.4	95.1	83.4	97.7	92.8	91.1	96.3	93.8	75.1	76.4
98.9	98.8	88.1	100.0	96.4	96.2	98.8	97.5	79.3	78.2
100.0	100.0	95.3	100.0	97.6	98.7	100.0	100.0	83.9	82.1
100.0	100.0	100.0	100.0	100.0	100.0	100.0	100.0	89.0	87.2
100.0	100.0	100.0	100.0	100.0	100.0	100.0	100.0	90.4	87.2
100.0	100.0	100.0	100.0	100.0	100.0	100.0	100.0	100.0	100.0

Table 1.3. Cumulative percentages: Adobe and Bedrock in Core 84

sample 1	sample 3	sample 6	sample 8	sample 7	sample 5	sample 4	sample 2	Bedrock PC 84
0.0	0.0	0.0	0.0	0.0	0.0	0.0	0.0	2.0
0.0	0.0	0.0	0.0	0.0	0.0	0.0	0.0	4.8
4.4	7.3	5.0	0.0	4.6	2.2	0.7	0.0	9.7
8.8	7.6	5.0	1.1	4.8	3.3	0.9	0.0	21.0
10.3	7.6	5.0	1.4	4.9	3.6	0.9	0.0	37.0
20.1	27.4	15.0	15.3	12.5	18.3	28.7	17.8	39.9
30.5	41.8	21.6	29.2	21.1	30.9	39.4	30.7	44.9
39.9	55.6	26.8	41.2	28.2	38.5	52.8	42.3	48.8
49.1	65.9	32.5	51.9	37.1	46.1	63.3	53.2	53.8
59.4	77.1	38.2	61.6	45.6	54.4	71.7	66.6	59.2
68.1	90.0	43.9	69.9	54.9	61.3	80.1	78.7	64.3
74.7	96.1	48.2	75.5	66.4	69.1	85.9	85.2	68.5
83.6	97.6	57.7	84.7	78.5	78.2	92.3	91.1	74.9
84.1	97.8	58.2	84.7	78.8	78.5	92.6	91.3	76.5
86.5	98.5	62.5	88.0	80.9	81.0	94.1	92.6	78.6
93.3	90.2	73.9	90.8	82.1	86.1	91.3	86.4	81.8
94.4	95.1	83.4	97.7	92.8	91.1	96.3	93.8	87.7
98.9	98.8	88.1	100.0	96.4	96.2	98.8	97.5	92.0
100.0	100.0	95.3	100.0	97.6	98.7	100.0	100.0	95.5
100.0	100.0	100.0	100.0	100.0	100.0	100.0	100.0	98.3
100.0	100.0	100.0	100.0	100.0	100.0	100.0	100.0	98.9
100.0	100.0	100.0	100.0	100.0	100.0	100.0	100.0	100.00

Table 1.4. Cumulative percentages: Adobe and Bedrock in Core 85

sample 1	sample 3	sample 6	sample 8	sample 7	sample 5	sample 4	sample 2	Bedrock PC 85
0.0	0.0	0.0	0.0	0.0	0.0	0.0	0.0	0.0
0.0	0.0	0.0	0.0	0.0	0.0	0.0	0.0	0.0
4.4	7.3	5.0	0.0	4.6	2.2	0.7	0.0	1.5
8.8	7.6	5.0	1.1	4.8	3.3	0.9	0.0	14.9
10.3	7.6	5.0	1.4	4.9	3.6	0.9	0.0	31.9
20.1	27.4	15.0	15.3	12.5	18.3	28.7	17.8	33.7
30.5	41.8	21.6	29.2	21.1	30.9	39.4	30.7	35.8
39.9	55.6	26.8	41.2	28.2	38.5	52.8	42.3	38.0
49.1	65.9	32.5	51.9	37.1	46.1	63.3	53.2	41.0
59.4	77.1	38.2	61.6	45.6	54.4	71.7	66.6	44.6
68.1	90.0	43.9	69.9	54.9	61.3	80.1	78.7	48.7
74.7	96.1	48.2	75.5	66.4	69.1	85.9	85.2	53.0
83.6	97.6	57.7	84.7	78.5	78.2	92.3	91.1	60.6
84.1	97.8	58.2	84.7	78.8	78.5	92.6	91.3	62.1
86.5	98.5	62.5	88.0	80.9	81.0	94.1	92.6	63.8
93.3	90.2	73.9	90.8	82.1	86.1	91.3	86.4	71.2
94.4	95.1	83.4	97.7	92.8	91.1	96.3	93.8	77.1
98.9	98.8	88.1	100.0	96.4	96.2	98.8	97.5	79.9
100.0	100.0	95.3	100.0	97.6	98.7	100.0	100.0	84.3
100.0	100.0	100.0	100.0	100.0	100.0	100.0	100.0	89.8
100.0	100.0	100.0	100.0	100.0	100.0	100.0	100.0	91.5
100.0	100.0	100.0	100.0	100.0	100.0	100.0	100.0	100.0

Table 1.5. Weight percentages in adobe materials

	gravel	sand	silt	clay
sample 1	10.2	76.2	13.4	0.0
sample 3	7.6	90.9	1.4	0.01
sample 6	4.9	57.5	37.5	0.0
sample 8	1.4	86.5	12.0	0.0
sample 7	4.8	76.0	19.0	0.01
sample 5	3.5	77.4	18.9	0.02
sample 4	0.9	93.1	5.9	0.01
sample 2	0.0	92.5	7.4	0.01

Table 1.6. Cumulative percentages: Adobe vs. Occupation deposits in the three Cores (Avrg.)

sample 1	sample 3	sample 6	sample 8	sample 7	sample 5	sample 4	sample 2	PC84 Zone 2d	PC84 Zone 2c	PC84 Zone 2b	PC84 Zone 2a	PC85 Zone 2d	PC85 Zone 2c	PC85 Zone 2b	PC85 Zone 2a	PC83 Zone 2b	PC83 Zone 2a
0.0	0.0	0.0	0.0	0.0	0.0	0.0	0.0	19.5	0.8	8.5	13.5	0.0	0.0	0.0	0.0	9.0	0.0
0.0	0.0	0.0	0.0	0.0	0.0	0.0	0.0	24.2	5.9	16.8	16.8	0.0	2.2	0.0	0.0	15.6	0.3
4.4	7.3	5.0	0.0	4.6	2.2	0.7	0.0	26.9	9.5	18.7	17.9	1.1	3.3	1.9	1.2	19.8	2.1
8.8	7.6	5.0	1.1	4.8	3.3	0.9	0.0	29.7	13.8	23.2	20.2	4.3	5.5	3.8	4.1	22.1	5.0
10.3	7.6	5.0	1.4	4.9	3.6	0.9	0.0	32.4	18.4	29.2	23.0	10.6	8.3	8.3	9.0	24.5	7.4
20.1	27.4	15.0	15.3	12.5	18.3	28.7	17.8	33.9	20.8	30.8	24.7	12.4	9.6	10.1	11.7	26.3	9.4

Table 1.6. (Continued)

sample 1	sample 3	sample 6	sample 8	sample 7	sample 5	sample 4	sample 2	PC84 Zone 2d	PC84 Zone 2c	PC84 Zone 2b	PC84 Zone 2a	PC85 Zone 2d	PC85 Zone 2c	PC85 Zone 2b	PC85 Zone 2a	PC83 Zone 2b	PC83 Zone 2a
30.5	41.8	21.6	29.2	21.1	30.9	39.4	30.7	35.8	23.4	32.6	26.6	14.5	11.3	13.2	16.0	27.8	11.9
39.9	55.6	26.8	41.2	28.2	38.5	52.8	42.3	38.0	26.8	34.8	29.0	16.7	13.5	16.9	21.0	29.5	14.8
49.1	65.9	32.5	51.9	37.1	46.1	63.3	53.2	40.6	30.8	37.6	32.2	19.4	16.4	21.4	26.8	31.8	18.2
59.4	77.1	38.2	61.6	45.6	54.4	71.7	66.6	43.4	35.1	40.8	35.8	22.4	19.6	26.2	32.4	34.7	22.0
68.1	90.0	43.9	69.9	54.9	61.3	80.1	78.7	46.6	39.9	44.4	40.1	26.3	23.3	31.5	38.2	38.5	26.1
74.7	96.1	48.2	75.5	66.4	69.1	85.9	85.2	49.6	44.4	47.8	44.2	30.1	26.9	36.4	42.9	42.1	29.7
83.6	97.6	57.7	84.7	78.5	78.2	92.3	91.1	52.4	48.3	51.2	48.6	33.9	30.5	41.3	47.1	44.8	33.3
84.1	97.8	58.2	84.7	78.8	78.5	92.6	91.3	54.3	51.0	54.0	51.8	37.1	33.5	45.0	50.7	47.0	36.1
86.5	98.5	62.5	88.0	80.9	81.0	94.1	92.6	55.9	53.2	56.4	54.7	39.8	35.9	48.3	54.0	48.8	38.3
93.3	90.2	73.9	90.8	82.1	86.1	91.3	86.4	72.9	72.1	74.1	72.8	59.5	58.8	73.3	78.4	69.0	61.2
94.4	95.1	83.4	97.7	92.8	91.1	96.3	93.8	78.4	78.5	80.3	77.9	68.4	66.8	79.4	82.9	74.0	68.4
98.9	98.8	88.1	100.0	96.4	96.2	98.8	97.5	82.3	83.7	84.4	83.5	76.3	73.1	84.0	87.5	78.2	75.5
100.0	100.0	95.3	100.0	97.6	98.7	100.0	100.0	84.8	86.1	87.5	87.3	82.0	77.7	88.6	89.8	82.2	80.5
100.0	100.0	100.0	100.0	100.0	100.0	100.0	100.0	88.2	88.4	90.5	90.9	88.7	82.3	90.1	92.0	85.5	85.4
100.0	100.0	100.0	100.0	100.0	100.0	100.0	100.0	90.2	92.8	93.0	93.6	94.3	88.0	93.9	95.5	89.3	90.3
100.0	100.0	100.0	100.0	100.0	100.0	100.0	100.0	100.0	100.0	100.0	100.0	100	100	100	100	100.00	100.0

Furthermore, when marrying sedimentation and sieve measurements, as in this study, the diameters of the hydrodynamically equivalent spheres, a ~ , obtained from the first technique must be converted to the sieve size obtained by the second. In general, this conversion requires knowledge, or estimation, of the shape of the particles. In the more realistic case that the shape of the particles is not known, as in this study, then this additive correction can be estimated from the data by incorporating it into the overall likelihood, replacing ci by ci+ E for those boundaries obtained by Stokes's law from the sedimentation technique. The simplicity of the basic log-Laplace model means that this is a viable technique, although care is needed in disentangling those sizes in the range covered by both techniques. The method is illustrated in Fieller and Flenley (1987) and the details are further discussed in Fieller and Flenley (1991).

Nevertheless, these parameters (i.e., mean, standard deviation, skewness and kyrtosis) are still presented by many workers despite their fundamental dependence on the unverified assumption of log-normality. As Gale and Hoare (1991: 64) state: *"their use is just acceptable so long as they are regarded simply as numerical indicators of a nature of a distribution which may be compared with those numerical indicators calculated by other workers"*. For this reason are also used here. In the present study ½ phi intervals were used in the calculations for detecting possible background "sedimental noise" on the grounds that "noise" may indicate some cultural activity which might otherwise (i.e., using whole phi) go unnoticed.

Gravel, sand, silt and clay weight percentages, cumulative percentages and summary statistics were calculated following the statistical procedures reported by Folk and Ward (1957) and Folk (1980). Only the cumulative percentages will be used in the present study and it should be noted that discrimination of differences between different cumulative frequency curves was established by eye.

1.3. APPLICATIONS OF P.S.A ON THE NEOLITHIC TELL

1.3a. Previous Applications of P.S.A on the Neolithic Tell

Initial human occupation on the Neolithic Tell site at Paliambela probably started on top of the Neogene sandy loam bedrock, because remnants of any overlying palaeosol have not been recognised, suggesting that this might have been stripped or reworked by subsequent human activity. Also, previous detailed textural analysis on occupation deposits from three Cores, drawn out from an extensive coring at the site, suggested that medium and coarse grained sands, small and large granule (2mm to 4mm) and small pebbles (4mm to 8mm) of these deposits may have been obtained from the available Neogene deposits underlying the site. Finally, comparative presentation of the particle size distribution between the gravelly muddy sand occupation deposits and the locally available alluvial sediments has shown that the site's deposits are similar in silt/clay content to the local alluvial deposits.

In short, the textural data indicate that the tell material was produced by introducing alluvial material and an additional contribution of coarser sands and finer gravel fractions (i.e., small granule to small pebbles) while a coarser component included in the texture of the occupation deposits, in the size of large (16mm to 32mm) pebbles, but unexplained by the texture of the natural sediments, was attributed to human activity, that is probably responsible for collecting, and carrying them to the site. Thus, the particle size analysis data led into the suggestion that the ca. 2m of accumulated occupation deposits may be attributed to introduced building material (for details see in Kontogiorgos, 2008).

1.3b. More Applications of P.SA. on the Neolithic Tell

The particle size analysis on adobe fragments provided the opportunity to link building material to tell formation. Comparative presentation of the particle size distribution between adobe materials, Holocene alluvial deposits from the area around and the Neogene sandy loams (i.e., bedrock) detected in the three Cores (i.e., PC 83,PC 84, PC 85) (Figures 1.2-1.5 and Tables 1.1-1.4) indicate that the adobe material from site's deposits is very similar in composition with the Neogene, sandy loam, bedrock. An interesting exception comes from adobe sample 6, which comprises more balanced composition in sand and silt particles in relation to the other adobe samples (Figure 1.6 and Table 1.5). Adobe sample 6 exhibits also compositional similarities with the occupation deposits from the Cores (Figure 1.7 and Table 1.6), but contrasts clearly with the other adobe samples. The implication of these results is that a significant amount of tell material may be attributed to, for example, eroded, weathered, and/or discarded adobe material comprising either as a major constituent Neogene sandy loams or balanced mixtures of sand and silt particles. Therefore, it could be argued that adobe material contributed to tell formation, providing medium and coarse grained sands, small and large granule (2mm to 4mm) and small pebbles (4mm to 8mm), but also finer, silty particles while the primary source for this material may have been obtained from the area around the Neolithic site.

The Rise of a Neolithic Tell

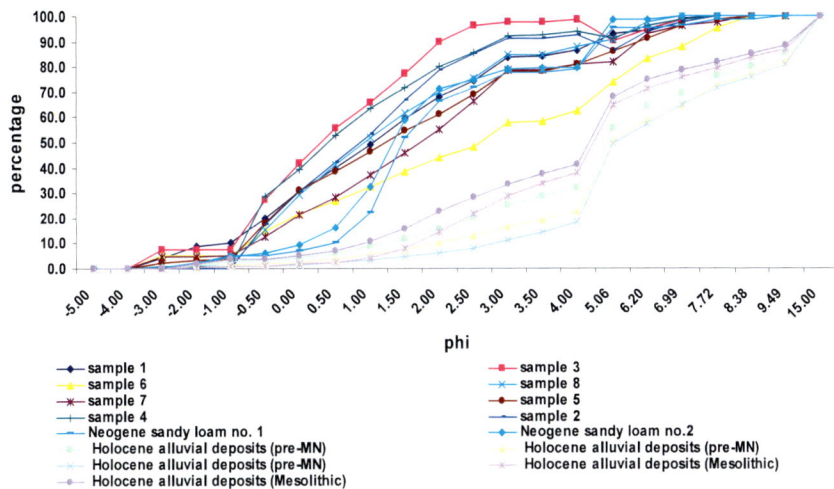

Figure 1.2. Cumulative frequency curves: Adobe (samples 1-8) vs. natural sediments (d.e.e.)[3].

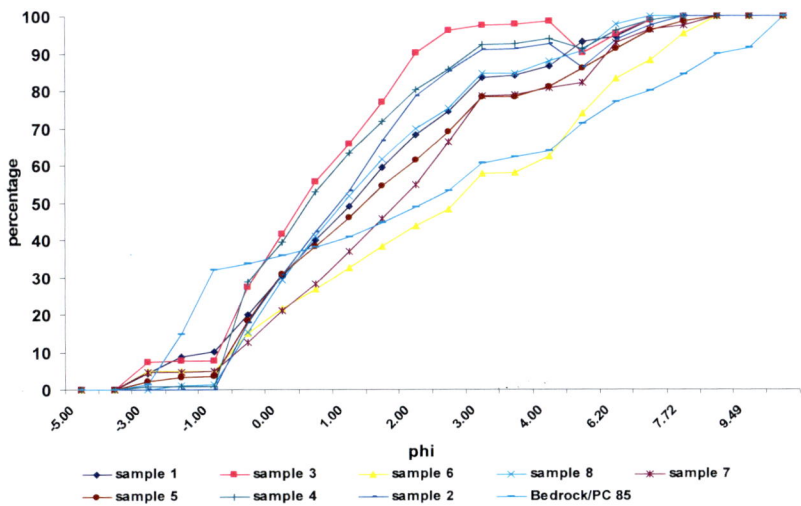

Figure 1.3. Cumulative frequency curves: Adobe vs. Bedrock in Core 85 (d.e.e.).

---

[3] discrimination established by eye.

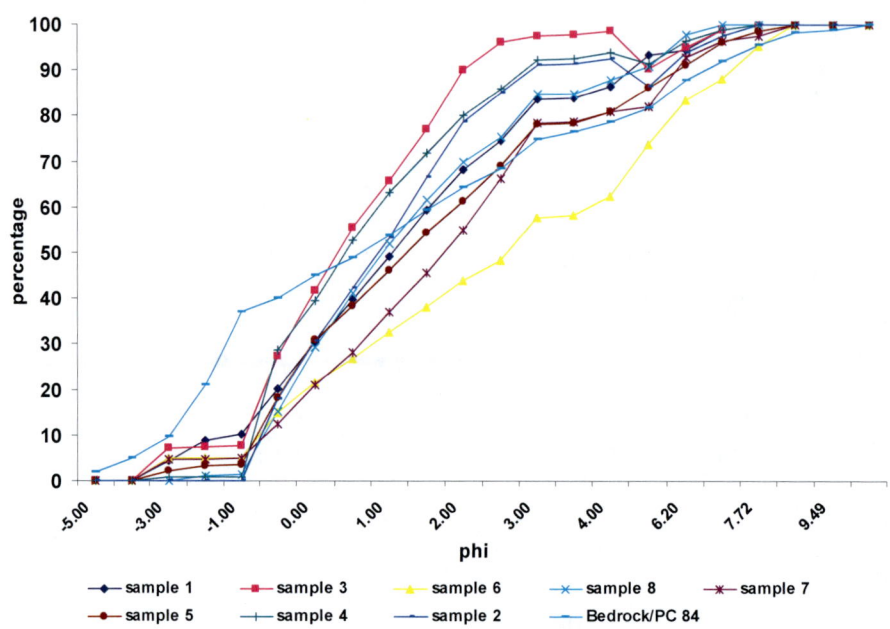

Figure 1.4. Cumulative frequency curves: Adobe vs. Bedrock in Core 84 (d.e.e.).

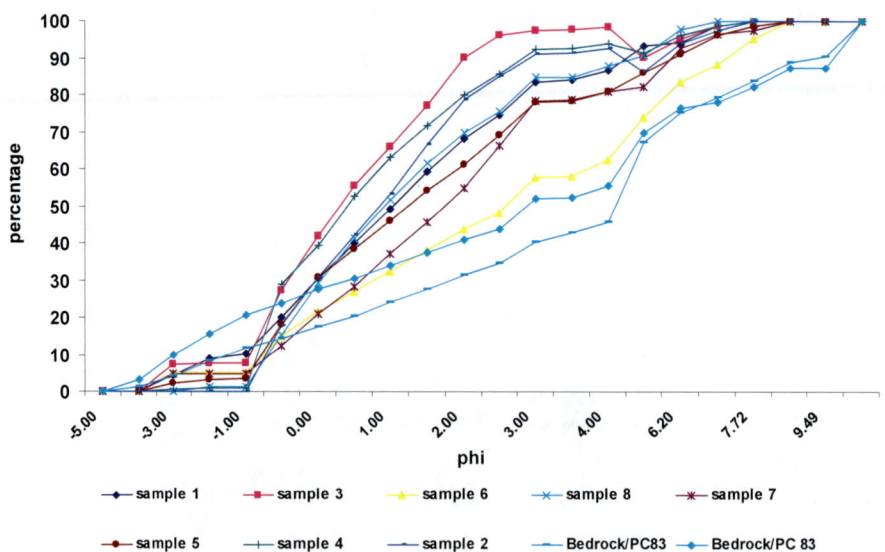

Figure 1.5. Cumulative frequency curves: Adobe vs. Bedrock in Core 83 (d.e.e.).

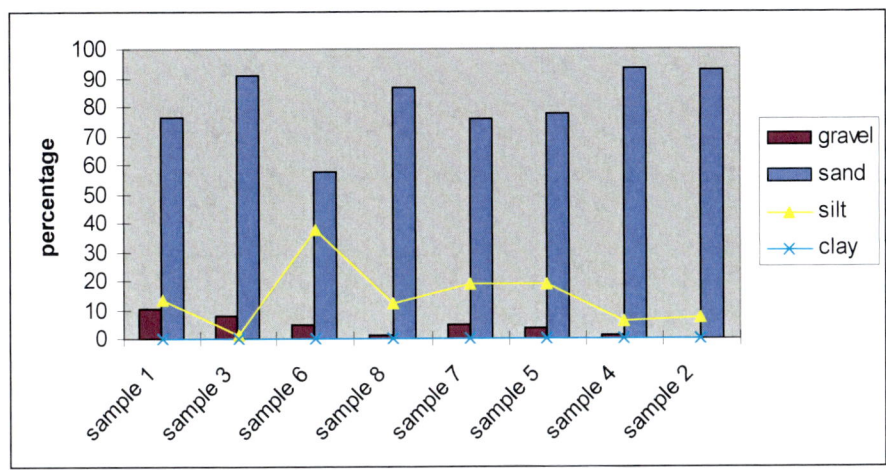

Figure 1.6. Particle size composition in adobe samples.

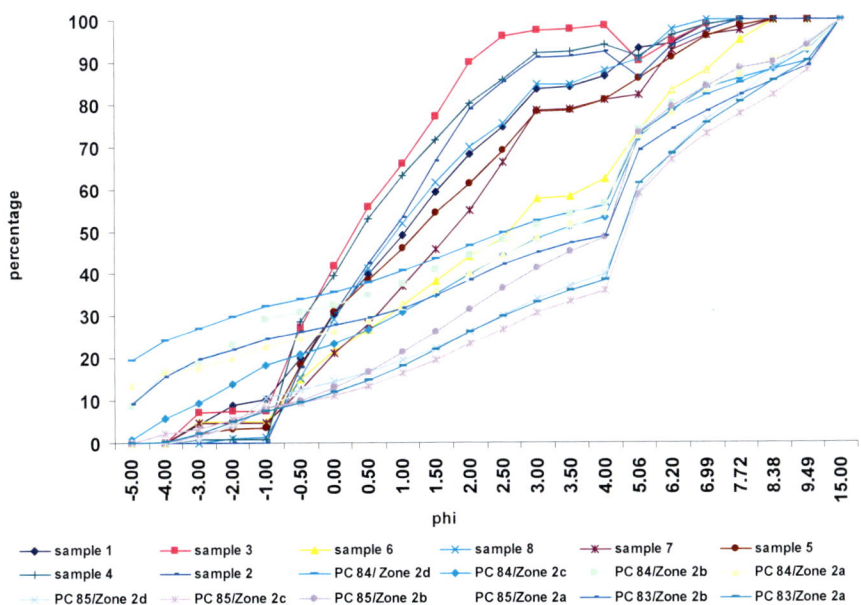

Figure 1.7. Cumulative frequency curves: Adobe vs. Occupation deposits in the three Cores (d.e.e.).

1.4. CONCLUSIONS

Particle size analysis, besides its drawbacks, can provide some insights into archaeological enquiries. Past and current applications of the method on the archaeological sediments from the Neolithic Tell site at Paliambela in Northern Greece, offered the opportunity to explore issues such as the sources of tell material and the contribution of building material on site formation. The current example from this Neolithic Tell in Northern Greece indicates significant contribution of sedimentary particles derived possibly from different types of building materials, in its formation. The primary source of these building materials and of the archaeological sediments (i.e., occupation deposits from Cores) may be located in the natural sediments around the site.

REFERENCES

ASTM (American Society for Testing Materials) (1961). Tentative method for grain-size analysis in soils. In *The 1961 Book of ASTM Standards*, pt.4, pp.1272-1283.

Barndorff-Nielsen, 0. (1977). Exponentially decreasing distributions for the logarithm of particle size. *Proc. R. Soc. A*, 353, pp. 401-419.

Bagnold, R. A. and Barndorff-Nielsen, 0. (1980). The pattern of natural size distributions. *Sedimentology*, 27, pp. 199-207.

Chapman, J. (1989). The early Balkan Village, In S. Bökönyi (ed.) *Neolithic of Southeastern Europe and its Near Eastern Connections*, pp. 33-53. Budapest: Varia Archaeologica Hungarica.

Davidson, D.A. (1973) Particle Size and Phoshate Analysis. Evidence for the Evolution of a Tell. *Archaeometry* 15, pp.143-152.

Davidson, D.A. (1976). Processes of Tell Formation and Erosion. In Davidson, D.A., and Shackley, M.L, (eds.), *Geoarchaeology: Earth Science and the Past*, pp. 143-158.

Fieller, N. R. J., and Flenley, E. C. (1987). Statistical analysis of particle sizes and sediments. *Br. Archaeol. Res. Ser.*, 393, pp. 79-94.

Fieller, N. R. J., and Flenley, E. C (1991). The combination of particle size data from different measurement methods. *Research Report* 381, pp. 91. Department of Probability and Statistics, University of Sheffield.

Fieller, N. R. J., Flenley, E. C., and Olbricht, W. (1992). Statistics of particle size data. *Applied. Statistics.* 41, pp. 127-46.

Fieller, N. R. J., Gilbertson, D. D. and Olbricht, W. (1984) A new method for environmental analysis of particle size distribution data from shoreline sediments. *Nature,* 311,pp. 648-651.

Folk, R.L. and Ward, C.W. (1957). Brazos river bar: A study in the significance of grain size parameters. *Journal of Sedimentary Petrology* 27 (1), pp. 3-26.

Folk, R.L. (1980). *Petrology of Sedimentary Rocks.* Austin: Hemphil.

Gale, S.J., and Hoare, P.G. (1991). *Quaternary Sediments. Petrographic Methods for the Study of Unlithified Rocks.* Halsted Press.

Galehouse, J.S. (1971). Sedimentation analysis. In Carver, R.E. (ed.), *Procedures in sedimentary petrology.* Wiley, New York, pp. 69-94.

Halstead, P. (1999). Neighbours from Hell? The Household in Neolithic Greece. In Halstead, P. (ed.), *Neolithic Society in Greece,* pp. 77-95. Sheffield Studies in Aegean Archaeology 2, Sheffield Academic Press.

Kaddah, M.T. (1974). The hydrometer method for detailed particle-size analysis: 1. Graphical interpretation of hydrometer readings and test of the method. *Soil Science* 14 (2), pp. 102-108.

Kolmogorov, A. N. (1941). *Uber das logarithmisch Normale Verteilungsgesetz der Dimensionen der Teilchen bei Zerstuckelung.* Compt. Rend. (Dokl.) Acad. Sci. URSS, 31, pp. 99-101.

Kontogiorgos, D. (2008). Geoarchaeological and Microartifact Analysis of Archaeological Sediments. A Case study From a Neolithic Tell Site in Greece. Nova Science Publishers, Inc., New York.

Kotsakis, K. (1999). What Tells Can Tell: Social Space and Settlement in the Greek Neolithic. In Halstead, P. (ed.), *Neolithic Society in Greece,* pp. 66-76. Sheffield Studies in Aegean Archaeology 2, Sheffield Academic Press.

Krumbein, W.C. (1934). Size frequency distributions of sediments. *Journal of Sedimentary Petrology* 4, pp. 65-77.

Krumbein, W. C. and Pettijohn, F. J. (1938). *Manual of Sedimentary Petrology.* New York: Appleton-Century-Crofts. Lloyd, S., 1963, *Mounds of the Near East.* Edinburgh.

Olbricht, W. (1982). *Modern statistical analysis of ancient sand.* MSc Thesis. University of Sheffield.

Otto, G. H. (1939). A modified logarithmic probability graph for the interpretation of mechanical analyses of sediments. *Journal of Sedimentary Petrology,* 9, pp. 62-76.

Rogers, J.J.W. (1965). Reproducibility and significance of measurements of sedimentary size distributions. *Journal of Sedimentary Petrology* 35, pp. 722-732.

Rosen, A.M. (1986). *Cities of Clay: The Geoarchaeology of Tells*. University of Chicago Press, Chicago.

Sherratt, A., 1983, Early Agrarian Settlement in the Koros Region of the Great Hungarian Plain. *Acta Archaeologica Academiae Scientiarum Hungaricae* 35, pp. 155-169.

Stevanovic, M., and Tringham, R. (1997). The significance of Neolithic houses in the archaeological record of southeast Europe. In Garasanin, M., Tasic, N., Cermanovic – Kuzmanovic, A., Petrovic, P., Mikic, Z. and Ruzic, M. (eds.), *Antidoron Dragoslavo Srejovic*, pp.195-207. Belgrade: Centre for Archaeological Research.

Tanner, W. F. (1958). The zig-zag nature of type I and type IV curves. *Journal of Sedimentary Petrology* 28, pp. 372-375.

In: On Site Geoarchaeology on a Neolithic Tell Site ... ISBN 978-1-60741-366-0
Editor: D. Kontogiorgos © 2009 Nova Science Publishers, Inc.

Chapter 2

TELL THE STORY OF A DITCH. ADDITIONAL SEDIMENTARY AND MICROARTIFACTUAL EVIDENCE FOR THE USE OF SPACE ON A NEOLITHIC TELL

Dimitris Kontogiorgos
Department of Archaeology, University of Sheffield, Northgate House, Sheffield, U.K.

2.1. INTRODUCTION

The geoarchaeological work at the Neolithic site of Paliambela (Northern Pieria region, Northern Greece), which unusually comprises both a compact tell and a flat-extended site, was initiated in order to explore issues regarding the continuity or discontinuity of occupation on the site and the effects of this upon the preservation of the deposits. The geoarchaeological study of formation processes and spatial organization at Paliambela was based on analysis of ditch and pit deposits from the tell and non-tell components of the site and of cores from the tell component. Although the focus of the investigation was the Neolithic period, pits of Byzantine-Ottoman date were also analyzed to explore temporal, spatial and contextual variation in deposition patterns/formation processes (for details see in Kontogiorgos, 2008).

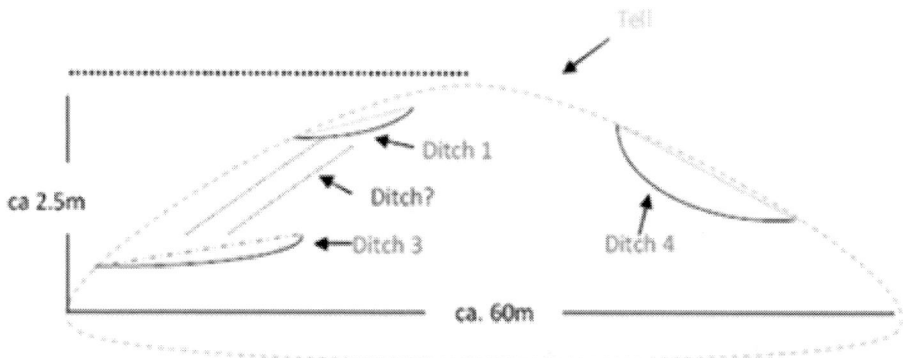

Figure 2.1. Location of the ditch on the Neolithic Tell (schematic, not in scale).

Specifically for the fills of the Neolithic ditches on the Tell site, these consisted of a rapid and probably deliberate primary fill attributable possibly to different forms of 'primary' depositional practices but also to differences in the amounts and types of excavated bedrock available to fall back into the ditches. A slower secondary upper fill was also detected in the ditches exhibiting marked variability through time in the modes of sedimentation and in the nature of cultural inputs suggesting differences in human activities occurring in the vicinity of the ditches.

This study complements the analysis of sediments and microartifacts of the Neolithic ditches from the Neolithic Tell at Paliambela by providing additional information from a deposit, tentatively interpreted as a ditch (Figure 2.1). The structure of this study is the following: section 2.2 describes the sampling procedure and the laboratory methods; section 2.3 presents the results of the geoarchaeological analysis and section 2.4 offers the conclusions of this study.

2.2. Sampling and Methodology

Macroscopic examination of the ditch fill clearly defined two basic stratigraphic units on the profile of the excavated context: bedrock and the ditch fill, a thick, coarse, brownish deposit (7.5YR 5/2), not so rich in cultural materials (Figure 2.2). It should be noted, however, that sampling was not proportional to the thickness and variability of the fill layers since macroscopic examination made clear that the layers were too complicated to accommodate such a sampling procedure.

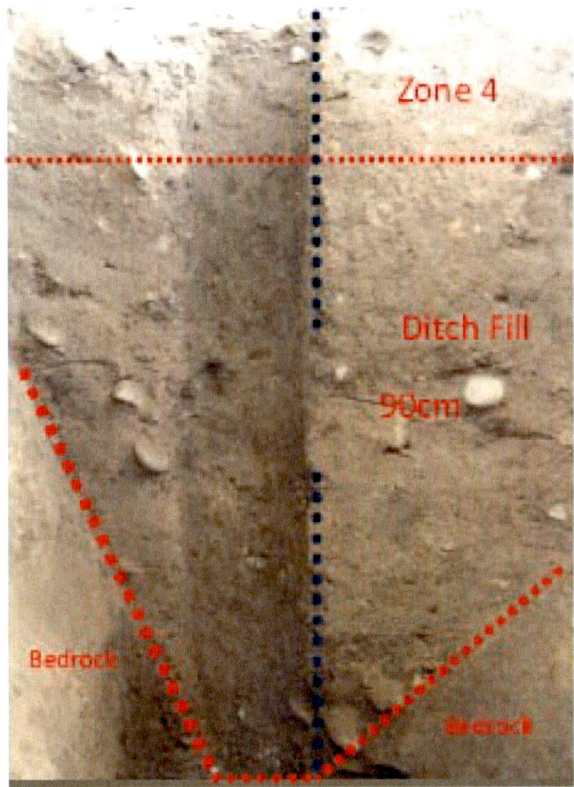

Source: Paliambela excavation archive.

Figure 2.2. The ditch after sampling.

A total of 18 sediment samples, with an average weight of ca 1500g, were collected in columns at 5cm vertical intervals on the profile of the context and were labeled according to depth. Two methods of analysis were applied in the sediment samples: particle size analysis for texture determination, and microartifact analysis for the cultural sedimentary particles smaller than 2mm in diameter (e.g., bone, shells, etc.). For the determination of particle size, hydrometer analysis (ASTM, 1961) was used for silt and clay, and sieving for the coarse particles (gravel and sand) (e.g., Folk, 1980) (Table 2.1). The procedure for determining the proportions of microartifact compositional types follows that described by Stein and Telster (1989: 10-11). One of the state-of-the-art minimization methodologies, the so-called Genetic Algorithms was applied for microartifact density determination (Kontogiorgos and Leontitsis, 2005; Chapter 3, this volume) (Table 2.2).

Table 2.1. Gravel, sand, silt and clay weight percentages for the ditch

Sample	gravel	sand	silt	clay
1	22.1	25.9	39.2	12.7
2	17.6	36.2	40.1	6.0
3	13.5	23.0	58.1	5.3
4	32.2	31.7	31.8	4.2
5	19.0	44.4	32.2	4.3
6	26.5	37.6	32.2	3.5
7	26.3	33.7	36.3	3.5
8	33.9	31.8	31.0	3.1
9	29.3	40.9	26.3	3.3
10	34.6	35.6	26.4	3.2
11	34.8	30.0	31.9	3.1
12	56.0	23.2	18.5	2.1
13	39.1	35.3	22.6	2.9
14	38.8	30.5	27.6	2.9
15	29.3	41.5	26.6	2.4
16	48.5	27.4	22.1	1.8
17	25.4	40.6	31.2	2.6
18	38.4	35.0	24.4	2.1

2.3. RESULTS AND INTERPRETATION (FIGURE 2.3)

Particle size analysis and microartifacts defined the stratigraphy of the Neolithic ditch. The lower part of the ditch (Zone 1) is the bedrock with low cultural inputs, probably intrusive from the upper layer (Zone 2). The lower part of the ditch fill (Zone 2) comprises variable values in gravel and sand, and very small quantities of microartifacts. The successive zone (Zone 3) again displays variability in sand and to a lesser extend gravel, while there is an increase in microartifacts values at the uppermost part of the zone. Peaks in shell and burnt clay at the upper part of Zone 3, however, co-vary with similar peak in sand. Finally, in Zone 4 there is an upwards increase in clay and shell concentration while a peak in silt at the bottom of this zone co-varies with peaks in bone and burnt clay.

Table 2.2. Microartifact density for the ditch

Sample	Shell	Bone	B.Clay
1	0,46%	0,00%	0,00%
2	0,87%	0,00%	0,01%
3	0,66%	0,01%	0,04%
4	0,10%	0,00%	0,00%
5	0,32%	0,00%	0,02%
6	0,20%	0,00%	0,00%
7	0,08%	0,00%	0,01%
8	0,00%	0,00%	0,00%
9	0,00%	0,00%	0,00%
10	0,07%	0,00%	0,00%
11	0,02%	0,00%	0,00%
12	0,12%	0,00%	0,00%
13	0,01%	0,00%	0,00%
14	0,00%	0,00%	0,00%
15	0,15%	0,00%	0,00%
16	0,12%	0,00%	0,00%
17	0,08%	0,00%	0,00%
18	0,18%	0,00%	0,00%

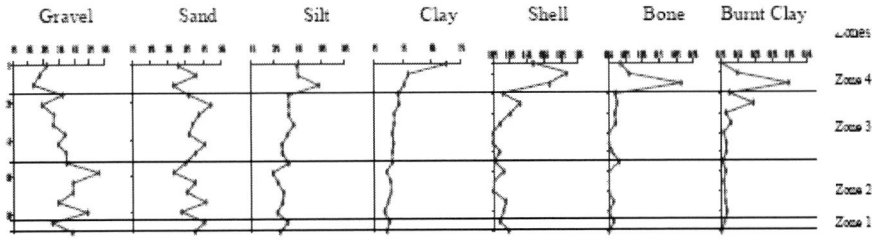

Figure 2.3. Particle size and microartifact data for the ditch.

The sedimentary and microartifactual characteristics of the analyzed deposit suggest rapid infilling due to fast sedimentation in Zones 2 and 3 without much cultural input. The increases in finer particles (silt and clay) in Zone 4 suggest slowing of sedimentation rates that might have triggered the accumulation of microartifacts at the upper part of the sequence. These formational characteristics are quite distinct from the ones observed in the three Neolithic ditches on the tell that exhibited abundance of microartifacts and variability in their values. Thus, while the deposit under examination displays relatively rapid accumulation, as was observed for the other Neolithic ditches on the Tell (Kontogiorgos, 2008) this may be largely attributed to fast sedimentation rather than to fast and/or variable cultural deposition. Therefore, the sedimentary and microartifactual characteristics for this deposit indicate different formation in relation to the other similar types of deposits on the tell and arguably imply different use of space at this part of the tell.

2.4. CONCLUSIONS

The specific sedimentary and artefactual characteristics of three Neolithic ditches on the tell site at Paliambela suggest spatially different activity areas in their vicinity (Kontogiorgos, 2008). The analysis of sediments and microartifacts from a deposit, tentatively interpreted as a ditch, demonstrates an archaeological feature exhibiting quite distinct formation from other similar types of contexts on the Tell. This result further denotes different use of this specific part of the Neolithic tell, while it complements the previous arguments for different activity areas nearby the ditches of the tell.

REFERENCES

ASTM (American Society for Testing Materials) (1961). Tentative method for grain-size analysis in soils. In *The 1961 Book of ASTM Standards*, pt.4, pp.1272-1283.

Folk, R.L. (1980). *Petrology of Sedimentary Rocks*. Austin: Hemphil.

Kontogiorgos, D, and Leontitsis, A. (2005). Micro-artefacts weight estimation by Genetic Algorithm minimisation, *Journal of Archaeological Science* 32(8), pp. 1275-1282.

Kontogiorgos, D. (2008). Geoarchaeological and Microartifact Analysis of Archaeological Sediments. A Case study From a Neolithic Tell Site in Greece. Nova Science Publishers, Inc.

Stein, J.K., and Telster, P.A. (1989). Size Distributions of Artifact Classes: Combining Macro- and Micro-Fractions. *Geoarchaeology*, 4(1), pp. 1-30.

In: On Site Geoarchaeology on a Neolithic Tell Site ... ISBN 978-1-60741-366-0
Editor: D. Kontogiorgos © 2009 Nova Science Publishers, Inc.

Chapter 3

TEST AGAIN FOR THE OPTIMUM! CONFIRMING THE USE OF GENETIC ALGORITHM MINIMIZATION ON MICROARTIFACT WEIGHT ESTIMATION

Dimitris Kontogiorgos[1], Alexandros Leontitsis[2,3]

[1] Department of Archaeology and Prehistory, University of Sheffield, Northgate House, Sheffield, U.K.

[2] Department of Culture and New Technologies, University of Ioannina, 45110 – Panepistimioupoli, Ioannina, Greece [e-mail: me00743@cc.uoi.gr]

[3] Center for Research and Applications of Nonlinear Systems, University of Patras, 26500 – Rio, Patras, Greece

3.1. INTRODUCTION

Several techniques for the quantification of microartifacts have been applied either by sorting and counting large number of samples (Metcalfe and Heath 1990; Simms and Heath, 1990); by estimating percentages of microartifact categories (Hassan, 1978; Rosen, 1989) or by point counting statistically representative samples (e.g., Stein and Telster, 1989; Sherwood and Ousley, 1995). A recent study (Kontogiorgos and Leontitsis, 2005) explored the use of Genetic Algorithms as an option for estimating the weights of microartifacts and the corresponding confidence intervals. The research has succeeded in coming with a consistent measure of the quantity of microartifacts from any single deposit, and thus could be used to obtain the proportional differences of the various microartifact classes between different sedimentary archaeological samples observing at the same time variations in the non-cultural sedimentary particles. The method achieved to estimate mean microartifact weight by minimizing the discrepancies caused due to the variation in the sedimentary structure of the deposits making in this way more precise possible comparisons with other types of geoarchaeological data such as particle size analysis data. In that previous attempt, the method was used experimentally and it was suggested that it should be tested in other cases since more applications were considered necessary to verify the proposed method.

This study tests the method once again and presents experimental microartifact data from different archaeological sites: a Neolithic Tell, an extended Neolithic settlement and a Hellenistic Theatre. The structure of this study is as follows: Section 2 briefly describes the concept and the technical details of Genetic Algorithms. Section 3 gives experimental results on archaeological data, and finally, section 4 presents the conclusions of this study.

3.2. AN OUTLINE OF GENETIC ALGORITHMS

Genetic Algorithms (GAs) are a class of stochastic algorithms widely used for solving optimisation problems. In fact, they are not solely stochastic but contain an underlying deterministic component. They lay somewhere in between pure chance and pure determinism, which gives them some of the advantages of both these two extremes (see, Schmitt (2001) for a discussion on the theoretical concepts of this particular class of model). The importance of Genetic Algorithms is obvious from the variety of their applications (Armano *et al.*, 2004; Csöndes *et al.*, 2002; Prügel-Bennet, 2004, among others). However, few variations have been proposed so far from the basic concept, which is mainly application-driven (i.e., Katayama *et al,*. 2001). Apart from the population size and the number of generations, they contain two basic parameters, which are crossover probability and mutation probability. These parameters are chosen when the algorithm is launched and remain constant throughout the whole process. Their values are decided by the researcher. Although experimentation on a particular application plays a very important role in their choice, crossover probability of 80%-100% and mutation probability of 1%-10% are more or less good choices.

If mutation probability is large, the whole algorithm will behave like a random search (i.e., will lose its deterministic properties), which is not desirable. Moreover, a small crossover probability will not add much to the evolution of the population, which is also not desirable. In this case the population members will remain unchanged as the generations evolve which means that the need for computational time will increase without any particular reason.

3.3. APPLICATIONS ON ARCHAEOLOGICAL DATA

The microartifact data presented below come from pits and a ditch of the Neolithic Tell site at Paliambela and the extended Neolithic settlement at Korinos, both located in the Northern Pieria region, Central Greek Macedonia, Northern Greece (e.g., Kotsakis and Halstead 2004, Besios and Adaktylou, 2004) and also, from colluvial deposits covering the auditorium of the Hellenistic theatre at Gitana, in Thesprotia region, Epirus, NW Greece (Kontogiorgos and Preka, Chapter 6, this volume).

A total of one hundred forty two (142) sediment samples were collected in columns at ca.10cm vertical intervals on the profiles of the Neolithic pits (i.e., pits nos. 1, 2, 3 and 4) and the ditch from the Neolithic tell, the EN pit 24, from the extended Neolithic settlement, as well as sediments from the colluvial deposit (i.e., GP1-GP5) [see Table 3.1, below]. The laboratory procedure used two divisions of the phi (ϕ) scale: - 2.00ϕ and 0ϕ. Contents of the bulk samples were passed through a stack of 4mm (-2.00ϕ) and 1mm (0ϕ) sieves. The material retained in the 1mm sieve created the sub-sample that was processed for microartifacts and an optical microscope was used for identifications. To avoid damaging the artifactual contents (e.g., shell, bone) there was no pretreatment for removal of organic matter or carbonate. The sub-sample was saturated with 1% sodium hexametaphoshate and washed through the 1mm sieve to separate the sand fraction from the silt/clay fraction. For each sub-sample, 500 particles were point-counted.

The identified microartifact types were: Microfragments of Burnt Clay, Microbone, Microshell, Microfragments of Charcoal, and natural sedimentary particles. To deal efficiently with the large numbers of samples derived from the contexts, and reduce the processing time, the point-counting procedure had to be applied. The procedure for determining the proportions of compositional types follows the one described by Stein and Telster (1989: 10-11). A small fraction of the sub-sample was poured gradually, into a glass petri-dish, below which was attached a piece of graph paper of no greater than 1cm graph intervals. The particles are spread evenly across the grid. While looking through the optical microscope, the particles located in one grid unit were counted according to compositional types. To improve the identifications and to observe more accurately the measurement error, 100 particles were counted and recorded each time, until a total of 500 particles were examined, since in the previous exercise a good stabilisation of the point estimation between 250 and 500 counted particles was achieved (Kontogiorgos and Leontitsis, 2005).

3.4. MICROARTIFACT MEAN WEIGHT ESTIMATION AND ESTIMATION OF CONFIDENCE INTERVALS

The first step was to construct a matrix **C** $m \times n$, with m rows and n columns. Every row contains the measurements for each sample, and every column contains the measurements for each material type: i.e., Burnt clay, Microbone, Microshell, Charcoal, and natural particles. Next, a column vector d with m elements was constructed that contains the sub-sample weights. The minimisation $\min_{x} \|\mathbf{Cx} - \mathbf{d}\|_2^2$ was introduced, under the constraint that every element of x is non-negative. The vector x is a column vector with n elements that represent the average particle weight of each material type.

A data-base was formed according to the above methodology and contains the measurements of the 500 particles (Table 3.1). The parameters m and n were 142 and 15, respectively. The parameters for both cases were the following: population 50, crossover probability 100%, mutation probability 0% increasing by 1% if there was no improvement on the global minimum, rank selection, decimal encoding with 8 digits, and the best 6 performing population members were maintained to breed the next generation (see De Falco *et al.* (2002) for experimentation with variations of the mutation process). In all cases, the obtained results on the 500 counted particles were used for the interpretation.

Table 3.1. Microartifact data base, containing the measurements on 500 particles
(microartifacts and natural sedimentary particles)

Total weight	Sample weight	Shell	Bone	Burnt Clay	Char-coal	pit 1	ditch	pit 2	pit 3	EN pit 24	pit 4	G.P 1	G.P 2	G.P 3	G.P 4	G.P 5	Esti-mation
881.3	136.8	13	4	51	11	421	0	0	0	0	0	0	0	0	0	0	168.71
1040.1	190.0	14	6	32	0	448	0	0	0	0	0	0	0	0	0	0	179.54
1279.2	133.3	11	2	48	0	439	0	0	0	0	0	0	0	0	0	0	175.78
1345.3	250.3	10	1	52	0	437	0	0	0	0	0	0	0	0	0	0	174.92
1132.0	242.6	3	0	50	0	447	0	0	0	0	0	0	0	0	0	0	178.51
1071.1	221.9	6	5	36	0	453	0	0	0	0	0	0	0	0	0	0	181.07
1167.2	280.0	16	0	45	0	449	0	0	0	0	0	0	0	0	0	0	180.05
1022.0	273.6	9	4	27	2	452	0	0	0	0	0	0	0	0	0	0	180.84
653.5	150.0	4	6	12	0	478	0	0	0	0	0	0	0	0	0	0	190.93
302.5	26.0	0	0	0	0	500	0	0	0	0	0	0	0	0	0	0	199.47
912.3	61.7	5	2	0	0	0	468	0	0	0	0	0	0	0	0	0	232.76
1344.1	104.6	16	6	32	0	0	446	0	0	0	0	0	0	0	0	0	222.47
1335.3	68.8	8	14	88	0	0	390	0	0	0	0	0	0	0	0	0	194.22
1487.4	329.3	6	6	52	0	0	436	0	0	0	0	0	0	0	0	0	216.94
1630.0	126.1	7	3	77	0	0	413	0	0	0	0	0	0	0	0	0	205.58
1694.3	344.2	12	6	30	0	0	452	0	0	0	0	0	0	0	0	0	225.22
1709.9	204.2	3	4	39	0	0	454	0	0	0	0	0	0	0	0	0	225.70
1710.0	458.1	0	0	25	0	0	475	0	0	0	0	0	0	0	0	0	235.96

Table 3.1. Continued

Total weight	Sample weight	Shell	Bone	Burnt Clay	Char-coal	pit 1	ditch	pit 2	pit 3	EN pit 24	pit 4	G.P 1	G.P 2	G.P 3	G.P 4	G.P 5	Esti-mation
1636.0	255.9	0	0	26	0	0	474	0	0	0	0	0	0	0	0	0	235.46
1196.8	237.3	3	2	24	0	0	469	0	0	0	0	0	0	0	0	0	233.15
1406.3	322.2	1	9	12	0	0	478	0	0	0	0	0	0	0	0	0	237.50
963.7	235.2	5	0	20	0	0	475	0	0	0	0	0	0	0	0	0	236.24
1534.3	439.0	1	0	23	0	0	476	0	0	0	0	0	0	0	0	0	236.51
1365.4	289.7	0	0	20	0	0	480	0	0	0	0	0	0	0	0	0	238.44
1023.0	235.4	6	3	19	0	0	472	0	0	0	0	0	0	0	0	0	234.81
1453.2	189.9	4	0	20	0	0	476	0	0	0	0	0	0	0	0	0	236.68
1177.6	150.2	2	2	10	0	0	486	0	0	0	0	0	0	0	0	0	241.53
416.6	64.9	2	0	0	0	0	498	0	0	0	0	0	0	0	0	0	247.49
1122.1	40.9	51	3	50	15	0	0	479	0	0	0	0	0	0	0	0	46.17
1038.1	36.1	39	13	42	10	0	0	399	0	0	0	0	0	0	0	0	38.27
1044.0	37.6	43	15	38	0	0	0	404	0	0	0	0	0	0	0	0	38.95
1191.2	38.7	53	10	29	15	0	0	403	0	0	0	0	0	0	0	0	39.42
1015.2	35.0	40	8	37	10	0	0	405	0	0	0	0	0	0	0	0	38.86
777.8	27.1	25	6	23	0	0	0	444	0	0	0	0	0	0	0	0	41.52
1118.5	68.8	30	5	45	15	0	0	405	0	0	0	0	0	0	0	0	38.29
1284.3	50.0	54	11	159	6	0	0	0	270	0	0	0	0	0	0	0	69.41
1739.4	65.4	50	15	146	3	0	0	0	286	0	0	0	0	0	0	0	73.10
2135.3	74.9	40	10	133	0	0	0	0	312	0	0	0	0	0	0	0	78.91
1290.1	56.8	59	6	170	8	0	0	0	257	0	0	0	0	0	0	0	66.50

1177.9	49.3	60	7	166	12	0	0	255	0	0	0	0	0	66.07
2263.2	89.2	54	10	156	10	0	0	270	0	0	0	0	0	69.41
1905.9	63.6	50	25	182	0	0	0	257	0	0	0	0	0	66.00
1583.5	48.5	35	15	192	20	0	0	233	0	0	0	0	0	59.25
1612.9	63.7	21	11	72	11	0	0	405	0	0	0	0	0	100.63
944.0	41.3	25	9	60	80	0	0	326	0	0	0	0	0	81.47
805.4	26.6	19	4	78	17	0	0	382	0	0	0	0	0	94.87
1189.1	38.9	18	4	42	7	0	0	429	0	0	0	0	0	106.34
1331.9	91.2	10	0	14	12	0	0	464	0	0	0	0	0	114.46
1585.6	101.6	0	0	4	2	0	0	494	0	0	0	0	0	121.25
1682.7	165.8	2	0	6	1	0	0	491	0	0	0	0	0	120.63
2259.1	234.8	4	0	11	1	0	0	484	0	0	0	0	0	119.03
1048.5	100.0	0	0	4	0	0	0	496	0	0	0	0	0	121.74
1162.9	129.4	1	1	2	0	0	0	496	0	0	0	0	0	121.80
1807.4	173.0	0	0	0	0	0	0	500	0	0	0	0	0	122.72
1457.6	136.3	0	0	0	0	0	0	500	0	0	0	0	0	122.72
2411.5	343.1	0	0	0	0	0	0	500	0	0	0	0	0	122.72
2557.0	267.0	0	0	0	0	0	0	500	0	0	0	0	0	122.72
909.9	34.3	0	0	0	0	0	0	500	0	0	0	0	0	122.72
1024.7	25.6	0	0	0	0	0	0	500	0	0	0	0	0	122.72
937.5	34.1	0	0	0	0	0	0	500	0	0	0	0	0	122.72
1696.4	59.8	0	0	0	0	0	0	500	0	0	0	0	0	122.72
1465.6	23.9	0	29	60	0	0	0	0	407	0	0	0	0	35.65
1440.3	38.1	17	25	99	0	0	0	0	359	0	0	0	0	32.43

Table 3.1. Continued

Total weight	Sample weight	Shel 1	Bone	Burnt Clay	Char-coal	pit 1	ditch	pit 2	pit 3	EN pit 24	pit 4	G.P 1	G.P 2	G.P 3	G.P 4	G.P 5	Esti-mation
1839.0	48.7	25	19	25	0	0	0	0	0	431	0	0	0	0	0	0	39.16
1113.0	29.2	25	19	25	0	0	0	0	0	431	0	0	0	0	0	0	39.16
847.3	33.5	16	7	20	0	0	0	0	0	457	0	0	0	0	0	0	40.91
864.0	18.0	10	10	40	0	0	0	0	0	440	0	0	0	0	0	0	39.09
969.5	20.9	10	24	50	0	0	0	0	0	416	0	0	0	0	0	0	37.00
1043.1	17.5	10	24	50	0	0	0	0	0	416	0	0	0	0	0	0	37.00
1439.9	26.1	11	60	40	0	0	0	0	0	389	0	0	0	0	0	0	34.71
1336.2	27.8	7	20	20	0	0	0	0	0	453	0	0	0	0	0	0	40.06
2022.9	47.5	4	22	35	0	0	0	0	0	439	0	0	0	0	0	0	38.66
1776.0	40.1	7	48	25	0	0	0	0	0	420	0	0	0	0	0	0	37.18
1500.1	14.0	9	12	27	0	0	0	0	0	450	0	0	0	0	0	0	39.91
1483.5	16.6	0	7	14	0	0	0	0	0	479	0	0	0	0	0	0	41.92
1331.3	20.4	3	3	10	0	0	0	0	0	484	0	0	0	0	0	0	42.53
1892.6	41.7	6	4	14	0	0	0	0	0	473	0	0	0	0	0	0	41.74
2030.0	64.6	6	11	14	0	0	0	0	0	442	0	0	0	0	0	0	39.03
1377.1	46.9	5	12	20	0	0	0	0	0	416	0	0	0	0	0	0	36.70
2127.1	98.6	5	12	20	0	0	0	0	0	416	0	0	0	0	0	0	36.70
1223.1	35.1	2	46	31	0	0	0	0	0	421	0	0	0	0	0	0	36.98
1505.2	33.6	0	35	60	0	0	0	0	0	405	0	0	0	0	0	0	35.47
1640.5	48.0	6	39	21	0	0	0	0	0	434	0	0	0	0	0	0	38.34
1157.2	94.1	5	0	23	0	0	0	0	0	472	0	0	0	0	0	0	41.60

141.7	5.1	62	40	100	0	0	0	0	472	0	0	0	0	5.47
195.0	7.9	58	60	90	0	0	0	0	292	0	0	0	0	4.53
188.9	0.5	54	20	80	0	0	0	0	346	0	0	0	0	4.50
186.7	0.9	62	10	100	0	0	0	0	328	0	0	0	0	4.88
190.7	0.8	22	10	100	0	0	0	0	368	0	0	0	0	2.76
167.2	0.5	20	15	60	0	0	0	0	405	0	0	0	0	2.78
183.7	0.9	24	80	40	0	0	0	0	356	0	0	0	0	2.84
140.0	3.2	15	40	20	0	0	0	0	425	0	0	0	0	2.58
180.2	1.3	4	10	10	0	0	0	0	476	0	0	0	0	2.14
173.2	5.0	0	0	0	0	0	0	0	500	0	0	0	0	2.00
132.0	3.9	0	0	0	0	0	0	0	500	0	0	0	0	2.00
192.0	4.8	0	0	0	0	0	0	0	500	0	0	0	0	2.00
185.9	7.3	0	0	0	0	0	0	0	500	0	0	0	0	2.00
1330.6	16.1	5	7	8	0	0	0	0	0	480	0	0	0	0.29
1563.9	16.6	12	10	48	0	0	0	0	0	430	0	0	0	0.70
1272.5	15.5	5	9	38	0	0	0	0	0	448	0	0	0	0.30
1888.0	12.9	4	5	15	0	0	0	0	0	476	0	0	0	0.24
1417.7	23.1	0	0	0	0	0	0	0	0	500	0	0	0	0.00
1090.3	5.3	15	10	29	0	0	0	0	0	0	446	0	0	25.65
789.9	8.7	20	12	69	0	0	0	0	0	0	399	0	0	23.34
934.8	14.3	30	18	56	0	0	0	0	0	0	396	0	0	23.74
1034.4	9.3	24	16	45	0	0	0	0	0	0	415	0	0	24.45
896.7	5.4	20	15	40	0	0	0	0	0	0	425	0	0	24.78
978.9	5.8	13	11	28	0	0	0	0	0	0	448	0	0	25.65
779.2	8.4	11	9	15	0	0	0	0	0	0	465	0	0	26.48

Table 3.1. Continued

Total weight	Sample weight	Shell	Bone	Burnt Clay	Char-coal	pit 1	ditch	pit 2	pit 3	EN pit 24	pit 4	G.P 1	G.P 2	G.P 3	G.P 4	G.P 5	Estimation
1205.7	30.1	5	6	18	0	0	0	0	0	0	0	0	471	0	0	0	26.47
1313.0	44.3	5	7	8	0	0	0	0	0	0	0	0	480	0	0	0	26.96
1387.1	64.4	0	0	0	0	0	0	0	0	0	0	0	500	0	0	0	27.78
951.9	6.1	5	7	8	0	0	0	0	0	0	0	0	0	480	0	0	7.47
1279.2	7.4	12	10	48	0	0	0	0	0	0	0	0	0	430	0	0	7.14
1904.6	8.6	5	9	38	0	0	0	0	0	0	0	0	0	448	0	0	7.01
1101.4	6.7	4	5	15	0	0	0	0	0	0	0	0	0	476	0	0	7.36
962.8	7.2	3	7	17	0	0	0	0	0	0	0	0	0	473	0	0	7.26
959.4	5.6	15	10	29	0	0	0	0	0	0	0	0	0	446	0	0	7.54
1279.1	8.1	20	12	69	0	0	0	0	0	0	0	0	0	399	0	0	7.14
1101.6	13.7	30	18	56	0	0	0	0	0	0	0	0	0	396	0	0	7.66
1530.4	18.2	24	16	45	0	0	0	0	0	0	0	0	0	415	0	0	7.60
677.0	12.2	20	15	40	0	0	0	0	0	0	0	0	0	425	0	0	7.52
1752.5	58.7	0	0	0	0	0	0	0	0	0	0	0	0	500	0	0	7.48
766.5	7.5	5	9	38	0	0	0	0	0	0	0	0	0	0	448	0	7.33
1306.1	9.7	4	5	15	0	0	0	0	0	0	0	0	0	0	476	0	7.70
1228.2	6.2	3	7	17	0	0	0	0	0	0	0	0	0	0	473	0	7.60
1448.4	8.7	15	10	59	0	0	0	0	0	0	0	0	0	0	416	0	7.41
1108.4	9.1	20	12	45	0	0	0	0	0	0	0	0	0	0	422	0	7.78
1249.5	10.3	13	18	56	0	0	0	0	0	0	0	0	0	0	412	0	7.23
632.4	5.2	5	6	15	0	0	0	0	0	0	0	0	0	0	474	0	7.73
1132.6	5.5	4	5	10	0	0	0	0	0	0	0	0	0	0	481	0	7.78
708.2	7.0	0	0	0	0	0	0	0	0	0	0	0	0	0	500	0	7.85
1052.0	31.3	5	4	18	0	0	0	0	0	0	0	0	0	0	0	473	20.92
1781.3	21.5	15	9	38	0	0	0	0	0	0	0	0	0	0	0	448	20.41
898.8	10.1	4	5	14	0	0	0	0	0	0	0	0	0	0	0	477	21.04

792.5	4.2	3	7	18	0	0	0	0	0	0	0	0	0	472	20.77
975.1	14.2	15	10	69	0	0	0	0	0	0	0	0	0	406	18.59
924.9	19.6	20	12	45	0	0	0	0	0	0	0	0	0	423	19.61
1335.2	36.1	13	18	36	0	0	0	0	0	0	0	0	0	443	20.08
1066.4	42.1	5	6	15	0	0	0	0	0	0	0	0	0	474	20.97
1436.6	49.1	4	5	10	0	0	0	0	0	0	0	0	0	481	21.21

Table 3.2. Average particle weight for the microartifacts and natural sedimentary particles and the estimated confidence intervals

	95% Lower Bound	Estimation	95% Upper Bound
Shell	0.0000	0.0285	0.7019
Bone	0.0000	0.0002	0.5004
Burnt Clay	0.0000	0.0002	0.2354
Charcoal	0.0000	0.0000	0.5456
Sed./pit 1	0.1656	0.1995	0.2506
Sed./pit 3	0.2167	0.2484	0.2799
Sed./pit 2	0.0000	0.0451	0.1159
Sed./pit 4	0.0923	0.1227	0.1534
Sed./ENpit24	0.0073	0.0438	0.0804
Sed./ditch	0.0000	0.0020	0.0477
Sed./G.P1	0.0000	0.0000	0.0673
Sed./G.P2	0.0000	0.0278	0.0714
Sed./G.P3	0.0000	0.0075	0.0588
Sed./G.P4	0.0000	0.0078	0.0518
Sed./G.P5	0.0000	0.0218	0.0716

Table 3.3. Microartifact weight percentages. Data obtained from the data base in Table 3.1

Shell	Bone	Burnt Clay	Char-coal	pit 1	ditch	pit 2	pit 3	EN pit 24	pit 4	G.P 1	G.P 2	G.P 3	G.P 4	G.P 5
0.54%	0.00%	0.01%	0.00%	122.77%	0.00%	0.00%	0.00%	0.00%	0.00%	0.00%	0.00%	0.00%	0.00%	0.00%
0.42%	0.00%	0.01%	0.00%	94.07%	0.00%	0.00%	0.00%	0.00%	0.00%	0.00%	0.00%	0.00%	0.00%	0.00%
0.47%	0.00%	0.01%	0.00%	131.38%	0.00%	0.00%	0.00%	0.00%	0.00%	0.00%	0.00%	0.00%	0.00%	0.00%
0.23%	0.00%	0.01%	0.00%	69.65%	0.00%	0.00%	0.00%	0.00%	0.00%	0.00%	0.00%	0.00%	0.00%	0.00%
0.07%	0.00%	0.01%	0.00%	73.51%	0.00%	0.00%	0.00%	0.00%	0.00%	0.00%	0.00%	0.00%	0.00%	0.00%
0.15%	0.00%	0.00%	0.00%	81.44%	0.00%	0.00%	0.00%	0.00%	0.00%	0.00%	0.00%	0.00%	0.00%	0.00%
0.33%	0.00%	0.00%	0.00%	63.97%	0.00%	0.00%	0.00%	0.00%	0.00%	0.00%	0.00%	0.00%	0.00%	0.00%
0.19%	0.00%	0.00%	0.00%	65.91%	0.00%	0.00%	0.00%	0.00%	0.00%	0.00%	0.00%	0.00%	0.00%	0.00%
0.15%	0.00%	0.00%	0.00%	127.13%	0.00%	0.00%	0.00%	0.00%	0.00%	0.00%	0.00%	0.00%	0.00%	0.00%
0.00%	0.00%	0.00%	0.00%	767.19%	0.00%	0.00%	0.00%	0.00%	0.00%	0.00%	0.00%	0.00%	0.00%	0.00%
0.46%	0.00%	0.00%	0.00%	0.00%	376.78%	0.00%	0.00%	0.00%	0.00%	0.00%	0.00%	0.00%	0.00%	0.00%
0.87%	0.00%	0.01%	0.00%	0.00%	211.80%	0.00%	0.00%	0.00%	0.00%	0.00%	0.00%	0.00%	0.00%	0.00%
0.66%	0.01%	0.04%	0.00%	0.00%	281.58%	0.00%	0.00%	0.00%	0.00%	0.00%	0.00%	0.00%	0.00%	0.00%
0.10%	0.00%	0.00%	0.00%	0.00%	65.77%	0.00%	0.00%	0.00%	0.00%	0.00%	0.00%	0.00%	0.00%	0.00%
0.32%	0.00%	0.02%	0.00%	0.00%	162.69%	0.00%	0.00%	0.00%	0.00%	0.00%	0.00%	0.00%	0.00%	0.00%
0.20%	0.00%	0.00%	0.00%	0.00%	65.23%	0.00%	0.00%	0.00%	0.00%	0.00%	0.00%	0.00%	0.00%	0.00%
0.08%	0.00%	0.01%	0.00%	0.00%	110.44%	0.00%	0.00%	0.00%	0.00%	0.00%	0.00%	0.00%	0.00%	0.00%
0.00%	0.00%	0.00%	0.00%	0.00%	51.51%	0.00%	0.00%	0.00%	0.00%	0.00%	0.00%	0.00%	0.00%	0.00%
0.00%	0.00%	0.00%	0.00%	0.00%	92.01%	0.00%	0.00%	0.00%	0.00%	0.00%	0.00%	0.00%	0.00%	0.00%
0.07%	0.00%	0.00%	0.00%	0.00%	98.17%	0.00%	0.00%	0.00%	0.00%	0.00%	0.00%	0.00%	0.00%	0.00%
0.02%	0.00%	0.00%	0.00%	0.00%	73.69%	0.00%	0.00%	0.00%	0.00%	0.00%	0.00%	0.00%	0.00%	0.00%
0.12%	0.00%	0.00%	0.00%	0.00%	100.32%	0.00%	0.00%	0.00%	0.00%	0.00%	0.00%	0.00%	0.00%	0.00%

0.01%	0.00%	0.00%	0.00%	0.00%	0.00%	0.00%	0.00%	0.00%	0.00%	0.00%	0.00%	0.00%	0.00%	0.00%
0.00%	0.00%	0.00%	0.00%	0.00%	0.00%	0.00%	0.00%	0.00%	0.00%	0.00%	0.00%	0.00%	0.00%	0.00%
0.15%	0.00%	0.00%	0.00%	53.86%	0.00%	0.00%	0.00%	0.00%	0.00%	0.00%	0.00%	0.00%	0.00%	0.00%
0.12%	0.00%	0.00%	0.00%	82.30%	0.00%	0.00%	0.00%	0.00%	0.00%	0.00%	0.00%	0.00%	0.00%	0.00%
0.08%	0.00%	0.00%	0.00%	99.60%	0.00%	0.00%	0.00%	0.00%	0.00%	0.00%	0.00%	0.00%	0.00%	0.00%
0.18%	0.00%	0.00%	0.00%	124.51%	0.00%	0.00%	0.00%	0.00%	0.00%	0.00%	0.00%	0.00%	0.00%	0.00%
7.12%	0.00%	0.00%	0.00%	160.73%	0.00%	0.00%	0.00%	0.00%	0.00%	0.00%	0.00%	0.00%	0.00%	0.00%
6.17%	0.02%	0.04%	0.00%	381.16%	105.74%	0.00%	0.00%	0.00%	0.00%	0.00%	0.00%	0.00%	0.00%	0.00%
6.53%	0.02%	0.04%	0.00%	0.00%	99.79%	0.00%	0.00%	0.00%	0.00%	0.00%	0.00%	0.00%	0.00%	0.00%
7.82%	0.01%	0.03%	0.00%	0.00%	97.01%	0.00%	0.00%	0.00%	0.00%	0.00%	0.00%	0.00%	0.00%	0.00%
6.52%	0.01%	0.02%	0.00%	0.00%	94.02%	0.00%	0.00%	0.00%	0.00%	0.00%	0.00%	0.00%	0.00%	0.00%
5.27%	0.01%	0.03%	0.00%	0.00%	104.47%	0.00%	0.00%	0.00%	0.00%	0.00%	0.00%	0.00%	0.00%	0.00%
2.49%	0.01%	0.03%	0.00%	0.00%	147.92%	0.00%	0.00%	0.00%	0.00%	0.00%	0.00%	0.00%	0.00%	0.00%
6.16%	0.00%	0.02%	0.00%	0.00%	53.15%	0.00%	0.00%	0.00%	0.00%	0.00%	0.00%	0.00%	0.00%	0.00%
4.36%	0.01%	0.10%	0.00%	0.00%	0.00%	132.54%	0.00%	0.00%	0.00%	0.00%	0.00%	0.00%	0.00%	0.00%
3.05%	0.01%	0.07%	0.00%	0.00%	0.00%	107.34%	0.00%	0.00%	0.00%	0.00%	0.00%	0.00%	0.00%	0.00%
5.93%	0.01%	0.05%	0.00%	0.00%	0.00%	102.24%	0.00%	0.00%	0.00%	0.00%	0.00%	0.00%	0.00%	0.00%
6.95%	0.00%	0.09%	0.00%	0.00%	0.00%	111.06%	0.00%	0.00%	0.00%	0.00%	0.00%	0.00%	0.00%	0.00%
3.46%	0.01%	0.10%	0.00%	0.00%	0.00%	126.96%	0.00%	0.00%	0.00%	0.00%	0.00%	0.00%	0.00%	0.00%
4.49%	0.00%	0.05%	0.00%	0.00%	0.00%	74.30%	0.00%	0.00%	0.00%	0.00%	0.00%	0.00%	0.00%	0.00%
4.12%	0.02%	0.09%	0.00%	0.00%	0.00%	99.18%	0.00%	0.00%	0.00%	0.00%	0.00%	0.00%	0.00%	0.00%
1.88%	0.01%	0.12%	0.00%	0.00%	0.00%	117.92%	0.00%	0.00%	0.00%	0.00%	0.00%	0.00%	0.00%	0.00%
3.46%	0.01%	0.03%	0.00%	0.00%	0.00%	156.05%	0.00%	0.00%	0.00%	0.00%	0.00%	0.00%	0.00%	0.00%
4.08%	0.01%	0.04%	0.00%	0.00%	0.00%	193.74%	0.00%	0.00%	0.00%	0.00%	0.00%	0.00%	0.00%	0.00%
	0.01%	0.09%	0.00%	0.00%	0.00%	352.49%	0.00%	0.00%	0.00%	0.00%	0.00%	0.00%	0.00%	0.00%

Table 3.3. Continued

Shell	Bone	Burnt Clay	Char-coal	pit 1	ditch	pit 2	pit 3	EN pit 24	pit 4	G.P 1	G.P 2	G.P 3	G.P 4	G.P 5
2.64%	0.00%	0.03%	0.00%	0.00%	0.00%	0.00%	270.69%	0.00%	0.00%	0.00%	0.00%	0.00%	0.00%	0.00%
0.63%	0.00%	0.00%	0.00%	0.00%	0.00%	0.00%	124.88%	0.00%	0.00%	0.00%	0.00%	0.00%	0.00%	0.00%
0.00%	0.00%	0.00%	0.00%	0.00%	0.00%	0.00%	119.34%	0.00%	0.00%	0.00%	0.00%	0.00%	0.00%	0.00%
0.07%	0.00%	0.00%	0.00%	0.00%	0.00%	0.00%	72.69%	0.00%	0.00%	0.00%	0.00%	0.00%	0.00%	0.00%
0.10%	0.00%	0.00%	0.00%	0.00%	0.00%	0.00%	50.60%	0.00%	0.00%	0.00%	0.00%	0.00%	0.00%	0.00%
0.00%	0.00%	0.00%	0.00%	0.00%	0.00%	0.00%	121.74%	0.00%	0.00%	0.00%	0.00%	0.00%	0.00%	0.00%
0.04%	0.00%	0.00%	0.00%	0.00%	0.00%	0.00%	94.08%	0.00%	0.00%	0.00%	0.00%	0.00%	0.00%	0.00%
0.00%	0.00%	0.00%	0.00%	0.00%	0.00%	0.00%	70.94%	0.00%	0.00%	0.00%	0.00%	0.00%	0.00%	0.00%
0.00%	0.00%	0.00%	0.00%	0.00%	0.00%	0.00%	90.04%	0.00%	0.00%	0.00%	0.00%	0.00%	0.00%	0.00%
0.00%	0.00%	0.00%	0.00%	0.00%	0.00%	0.00%	35.77%	0.00%	0.00%	0.00%	0.00%	0.00%	0.00%	0.00%
0.00%	0.00%	0.00%	0.00%	0.00%	0.00%	0.00%	45.96%	0.00%	0.00%	0.00%	0.00%	0.00%	0.00%	0.00%
0.00%	0.00%	0.00%	0.00%	0.00%	0.00%	0.00%	357.80%	0.00%	0.00%	0.00%	0.00%	0.00%	0.00%	0.00%
0.00%	0.00%	0.00%	0.00%	0.00%	0.00%	0.00%	479.39%	0.00%	0.00%	0.00%	0.00%	0.00%	0.00%	0.00%
0.00%	0.00%	0.00%	0.00%	0.00%	0.00%	0.00%	359.90%	0.00%	0.00%	0.00%	0.00%	0.00%	0.00%	0.00%
0.00%	0.00%	0.00%	0.00%	0.00%	0.00%	0.00%	205.22%	0.00%	0.00%	0.00%	0.00%	0.00%	0.00%	0.00%
0.00%	0.05%	0.08%	0.00%	0.00%	0.00%	0.00%	0.00%	149.02%	0.00%	0.00%	0.00%	0.00%	0.00%	0.00%
2.55%	0.03%	0.08%	0.00%	0.00%	0.00%	0.00%	0.00%	82.46%	0.00%	0.00%	0.00%	0.00%	0.00%	0.00%
2.93%	0.02%	0.02%	0.00%	0.00%	0.00%	0.00%	0.00%	77.45%	0.00%	0.00%	0.00%	0.00%	0.00%	0.00%
4.89%	0.03%	0.03%	0.00%	0.00%	0.00%	0.00%	0.00%	129.17%	0.00%	0.00%	0.00%	0.00%	0.00%	0.00%
2.73%	0.01%	0.02%	0.00%	0.00%	0.00%	0.00%	0.00%	119.38%	0.00%	0.00%	0.00%	0.00%	0.00%	0.00%
3.17%	0.02%	0.07%	0.00%	0.00%	0.00%	0.00%	0.00%	213.91%	0.00%	0.00%	0.00%	0.00%	0.00%	0.00%
2.73%	0.05%	0.07%	0.00%	0.00%	0.00%	0.00%	0.00%	174.18%	0.00%	0.00%	0.00%	0.00%	0.00%	0.00%

3.26%	0.06%	0.09%	0.00%	0.00%	0.00%	0.00%	208.02%	0.00%	0.00%	0.00%	0.00%	0.00%	0.00%
2.41%	0.10%	0.05%	0.00%	0.00%	0.00%	0.00%	130.42%	0.00%	0.00%	0.00%	0.00%	0.00%	0.00%
1.44%	0.03%	0.02%	0.00%	0.00%	0.00%	0.00%	142.59%	0.00%	0.00%	0.00%	0.00%	0.00%	0.00%
0.48%	0.02%	0.02%	0.00%	0.00%	0.00%	0.00%	80.88%	0.00%	0.00%	0.00%	0.00%	0.00%	0.00%
1.00%	0.05%	0.02%	0.00%	0.00%	0.00%	0.00%	91.65%	0.00%	0.00%	0.00%	0.00%	0.00%	0.00%
3.67%	0.04%	0.06%	0.00%	0.00%	0.00%	0.00%	281.28%	0.00%	0.00%	0.00%	0.00%	0.00%	0.00%
0.00%	0.02%	0.03%	0.00%	0.00%	0.00%	0.00%	252.51%	0.00%	0.00%	0.00%	0.00%	0.00%	0.00%
0.84%	0.01%	0.02%	0.00%	0.00%	0.00%	0.00%	207.62%	0.00%	0.00%	0.00%	0.00%	0.00%	0.00%
0.82%	0.00%	0.01%	0.00%	0.00%	0.00%	0.00%	99.26%	0.00%	0.00%	0.00%	0.00%	0.00%	0.00%
0.53%	0.01%	0.01%	0.00%	0.00%	0.00%	0.00%	59.87%	0.00%	0.00%	0.00%	0.00%	0.00%	0.00%
0.61%	0.01%	0.01%	0.00%	0.00%	0.00%	0.00%	77.62%	0.00%	0.00%	0.00%	0.00%	0.00%	0.00%
0.29%	0.01%	0.01%	0.00%	0.00%	0.00%	0.00%	36.92%	0.00%	0.00%	0.00%	0.00%	0.00%	0.00%
0.33%	0.06%	0.03%	0.00%	0.00%	0.00%	0.00%	104.96%	0.00%	0.00%	0.00%	0.00%	0.00%	0.00%
0.00%	0.04%	0.05%	0.00%	0.00%	0.00%	0.00%	105.48%	0.00%	0.00%	0.00%	0.00%	0.00%	0.00%
0.71%	0.03%	0.01%	0.00%	0.00%	0.00%	0.00%	79.12%	0.00%	0.00%	0.00%	0.00%	0.00%	0.00%
0.30%	0.00%	0.01%	0.00%	0.00%	0.00%	0.00%	43.89%	0.00%	0.00%	0.00%	0.00%	0.00%	0.00%
6.39%	0.33%	0.60%	0.00%	0.00%	0.00%	0.00%	0.00%	36.94%	0.00%	0.00%	0.00%	0.00%	0.00%
4.91%	0.32%	0.35%	0.00%	0.00%	0.00%	0.00%	0.00%	14.75%	0.00%	0.00%	0.00%	0.00%	0.00%
5.78%	1.63%	4.74%	0.00%	0.00%	0.00%	0.00%	0.00%	265.55%	0.00%	0.00%	0.00%	0.00%	0.00%
3.50%	0.45%	3.28%	0.00%	0.00%	0.00%	0.00%	0.00%	139.26%	0.00%	0.00%	0.00%	0.00%	0.00%
1.24%	0.56%	4.05%	0.00%	0.00%	0.00%	0.00%	0.00%	193.24%	0.00%	0.00%	0.00%	0.00%	0.00%
2.55%	1.22%	3.55%	0.00%	0.00%	0.00%	0.00%	0.00%	310.83%	0.00%	0.00%	0.00%	0.00%	0.00%
1.17%	4.00%	1.45%	0.00%	0.00%	0.00%	0.00%	0.00%	167.15%	0.00%	0.00%	0.00%	0.00%	0.00%
2.76%	0.53%	0.19%	0.00%	0.00%	0.00%	0.00%	0.00%	53.00%	0.00%	0.00%	0.00%	0.00%	0.00%

Table 3.3. Continued

Shell	Bone	Burnt Clay	Char-coal	pit 1	ditch	pit 2	pit 3	EN pit 24	pit 4	G.P 1	G.P 2	G.P 3	G.P 4	G.P 5
1.56%	0.33%	0.24%	0.00%	0.00%	0.00%	0.00%	0.00%	0.00%	146.13%	0.00%	0.00%	0.00%	0.00%	0.00%
0.00%	0.00%	0.00%	0.00%	0.00%	0.00%	0.00%	0.00%	0.00%	39.28%	0.00%	0.00%	0.00%	0.00%	0.00%
0.00%	0.00%	0.00%	0.00%	0.00%	0.00%	0.00%	0.00%	0.00%	51.16%	0.00%	0.00%	0.00%	0.00%	0.00%
0.00%	0.00%	0.00%	0.00%	0.00%	0.00%	0.00%	0.00%	0.00%	41.40%	0.00%	0.00%	0.00%	0.00%	0.00%
0.00%	0.00%	0.00%	0.00%	0.00%	0.00%	0.00%	0.00%	0.00%	27.45%	0.00%	0.00%	0.00%	0.00%	0.00%
1.77%	0.02%	0.02%	0.00%	0.00%	0.00%	0.00%	0.00%	0.00%	0.00%	0.00%	0.00%	0.00%	0.00%	0.00%
4.13%	0.03%	0.09%	0.00%	0.00%	0.00%	0.00%	0.00%	0.00%	0.00%	0.00%	0.00%	0.00%	0.00%	0.00%
1.84%	0.02%	0.08%	0.00%	0.00%	0.00%	0.00%	0.00%	0.00%	0.00%	0.00%	0.00%	0.00%	0.00%	0.00%
1.77%	0.02%	0.04%	0.00%	0.00%	0.00%	0.00%	0.00%	0.00%	0.00%	0.00%	0.00%	0.00%	0.00%	0.00%
0.00%	0.00%	0.00%	0.00%	0.00%	0.00%	0.00%	0.00%	0.00%	0.00%	0.00%	0.00%	0.00%	0.00%	0.00%
16.16%	0.08%	0.17%	0.00%	0.00%	0.00%	0.00%	0.00%	0.00%	0.00%	0.00%	467.60%	0.00%	0.00%	0.00%
13.12%	0.06%	0.24%	0.00%	0.00%	0.00%	0.00%	0.00%	0.00%	0.00%	0.00%	254.84%	0.00%	0.00%	0.00%
11.98%	0.05%	0.12%	0.00%	0.00%	0.00%	0.00%	0.00%	0.00%	0.00%	0.00%	153.88%	0.00%	0.00%	0.00%
14.73%	0.07%	0.15%	0.00%	0.00%	0.00%	0.00%	0.00%	0.00%	0.00%	0.00%	247.96%	0.00%	0.00%	0.00%
21.14%	0.12%	0.23%	0.00%	0.00%	0.00%	0.00%	0.00%	0.00%	0.00%	0.00%	437.33%	0.00%	0.00%	0.00%
12.79%	0.08%	0.15%	0.00%	0.00%	0.00%	0.00%	0.00%	0.00%	0.00%	0.00%	429.21%	0.00%	0.00%	0.00%
7.48%	0.05%	0.05%	0.00%	0.00%	0.00%	0.00%	0.00%	0.00%	0.00%	0.00%	307.60%	0.00%	0.00%	0.00%
0.95%	0.01%	0.02%	0.00%	0.00%	0.00%	0.00%	0.00%	0.00%	0.00%	0.00%	86.95%	0.00%	0.00%	0.00%
0.64%	0.01%	0.01%	0.00%	0.00%	0.00%	0.00%	0.00%	0.00%	0.00%	0.00%	60.21%	0.00%	0.00%	0.00%
0.00%	0.00%	0.00%	0.00%	0.00%	0.00%	0.00%	0.00%	0.00%	0.00%	0.00%	43.14%	0.00%	0.00%	0.00%
4.68%	0.05%	0.04%	0.00%	0.00%	0.00%	0.00%	0.00%	0.00%	0.00%	0.00%	0.00%	117.76%	0.00%	0.00%
9.26%	0.06%	0.20%	0.00%	0.00%	0.00%	0.00%	0.00%	0.00%	0.00%	0.00%	0.00%	86.96%	0.00%	0.00%
3.32%	0.04%	0.14%	0.00%	0.00%	0.00%	0.00%	0.00%	0.00%	0.00%	0.00%	0.00%	77.96%	0.00%	0.00%
3.41%	0.03%	0.07%	0.00%	0.00%	0.00%	0.00%	0.00%	0.00%	0.00%	0.00%	0.00%	106.32%	0.00%	0.00%

2.38%	0.04%	0.07%	0.00%	0.00%	0.00%	0.00%	0.00%	0.00%	0.00%	98.32%	0.00%	0.00%
15.29%	0.08%	0.16%	0.00%	0.00%	0.00%	0.00%	0.00%	0.00%	0.00%	119.19%	0.00%	0.00%
14.09%	0.06%	0.26%	0.00%	0.00%	0.00%	0.00%	0.00%	0.00%	0.00%	73.72%	0.00%	0.00%
12.50%	0.06%	0.13%	0.00%	0.00%	0.00%	0.00%	0.00%	0.00%	0.00%	43.26%	0.00%	0.00%
7.53%	0.04%	0.08%	0.00%	0.00%	0.00%	0.00%	0.00%	0.00%	0.00%	34.13%	0.00%	0.00%
9.36%	0.05%	0.10%	0.00%	0.00%	0.00%	0.00%	0.00%	0.00%	0.00%	52.14%	0.00%	0.00%
0.00%	0.00%	0.00%	0.00%	0.00%	0.00%	0.00%	0.00%	0.00%	0.00%	12.75%	0.00%	0.00%
3.81%	0.05%	0.16%	0.00%	0.00%	0.00%	0.00%	0.00%	0.00%	0.00%	0.00%	93.74%	0.00%
2.35%	0.02%	0.05%	0.00%	0.00%	0.00%	0.00%	0.00%	0.00%	0.00%	0.00%	77.01%	0.00%
2.76%	0.05%	0.08%	0.00%	0.00%	0.00%	0.00%	0.00%	0.00%	0.00%	0.00%	119.72%	0.00%
9.84%	0.05%	0.21%	0.00%	0.00%	0.00%	0.00%	0.00%	0.00%	0.00%	0.00%	75.04%	0.00%
12.55%	0.06%	0.15%	0.00%	0.00%	0.00%	0.00%	0.00%	0.00%	0.00%	0.00%	72.77%	0.00%
7.20%	0.07%	0.17%	0.00%	0.00%	0.00%	0.00%	0.00%	0.00%	0.00%	0.00%	62.77%	0.00%
5.49%	0.05%	0.09%	0.00%	0.00%	0.00%	0.00%	0.00%	0.00%	0.00%	0.00%	143.04%	0.00%
4.15%	0.04%	0.06%	0.00%	0.00%	0.00%	0.00%	0.00%	0.00%	0.00%	0.00%	137.24%	0.00%
0.00%	0.00%	0.00%	0.00%	0.00%	0.00%	0.00%	0.00%	0.00%	0.00%	0.00%	112.09%	0.00%
0.91%	0.01%	0.02%	0.00%	0.00%	0.00%	0.00%	0.00%	0.00%	0.00%	0.00%	0.00%	65.91%
3.98%	0.02%	0.05%	0.00%	0.00%	0.00%	0.00%	0.00%	0.00%	0.00%	0.00%	0.00%	90.88%
2.26%	0.02%	0.04%	0.00%	0.00%	0.00%	0.00%	0.00%	0.00%	0.00%	0.00%	0.00%	205.99%
4.08%	0.07%	0.13%	0.00%	0.00%	0.00%	0.00%	0.00%	0.00%	0.00%	0.00%	0.00%	490.16%
6.03%	0.03%	0.15%	0.00%	0.00%	0.00%	0.00%	0.00%	0.00%	0.00%	0.00%	0.00%	124.71%
5.82%	0.03%	0.07%	0.00%	0.00%	0.00%	0.00%	0.00%	0.00%	0.00%	0.00%	0.00%	94.13%
2.06%	0.02%	0.03%	0.00%	0.00%	0.00%	0.00%	0.00%	0.00%	0.00%	0.00%	0.00%	53.52%
0.68%	0.01%	0.01%	0.00%	0.00%	0.00%	0.00%	0.00%	0.00%	0.00%	0.00%	0.00%	49.11%
0.47%	0.00%	0.01%	0.00%	0.00%	0.00%	0.00%	0.00%	0.00%	0.00%	0.00%	0.00%	42.73%

Table 3.2 presents the average particle weight for the microartifacts and natural sedimentary particles, as these were estimated through *minimisation*. As Table 3.2 shows, the estimated average weight per particle for the material types and the natural sedimentary particles present in the sequences is non-negative. Table 3.2 also shows the confidence intervals of the point estimates. More precisely, it demonstrates the individual confidence intervals of each material type. These confidence intervals can be obtained by the chi-squared test, which is based on the likelihood of the minimisation model. Let $\chi^2_{1,95\%}$ be the value of inverse chi-squared distribution with 1 degree of freedom and probability 95%. The upper and the lower bound should fulfil the equation $-m(\log(\hat{\mathbf{x}}) - \log(\mathbf{x}_0)) = \chi^2_{1,95\%}$, where $\hat{\mathbf{x}}$ stands for the value of x estimated by minimisation, and $\mathbf{x}_0$ is the perturbed x (Morgan, 2001). Finally, Table 3.3 shows the microartifacts weight percentages.

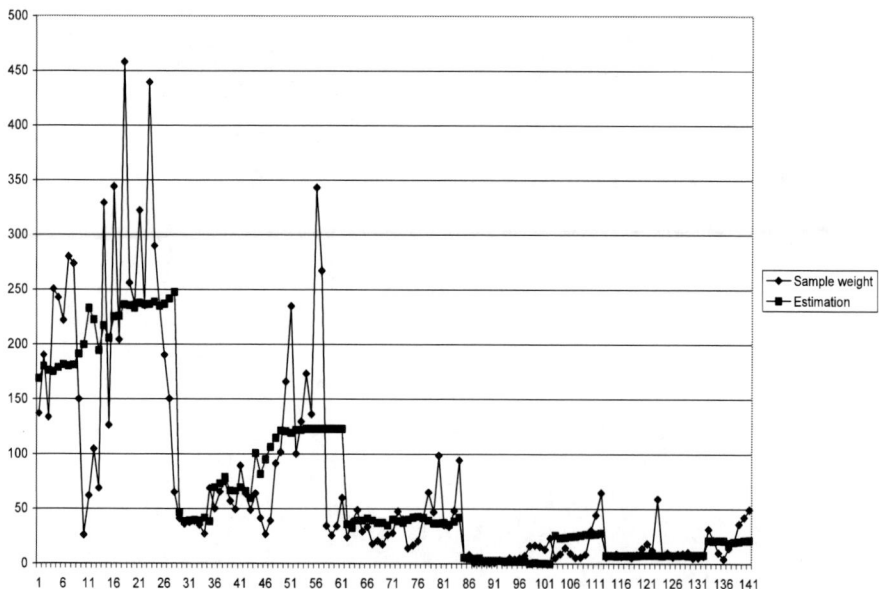

Figure 3.1. Discrepancies between actual sub-sample weight values and the estimated sub-sample weight values from the data obtained from the data-base.

Discrepancies may be observed between the actual sub-sample weight values (d) and the estimated values (Cx). These discrepancies have to be relatively small. Figure 3.1 presents the discrepancies between actual sub-sample weight values and the estimated sub-sample weight values from the data obtained from the database.

3.5. CONCLUSIONS

This study explored once again the use of Genetic Algorithms as an option for estimating the weights of microartifacts and the corresponding confidence intervals. The method achieved to estimate mean microartifact weight by minimizing the discrepancies caused may be due to the variation in the depositional and post-depositional disturbances on different types of archaeological sediments and contexts. Therefore, it potentially offers a tool to the researcher attempting a similar undertaken, to work with large data-bases including data from different types of contexts and thus saving time and effort cost.

REFERENCES

Armano, G., Marchesi, M., and Murru, A. (2004). A hybrid genetic-neural architecture for stock indixes forecasting. *Information Sciences*, to appear

Besios, M., and Adaktylou, F. (2004). Neolithikos Ikismos sta 'Ravenia' Korinou. *Paper presented at Archaeologikon Ergon sti Makedonia and sti Thraki Conference 2004.* (In Greek).

Csöndes, T., Kotnyek, B., and Szabó, J. Z. (2002). Application of heuristic methods for conformance test selection. *European Journal of Operations Research* 142, pp. 203-218.

De Falco, I., Della Cioppa, A., and Tarantino, E. (2002). Mutation-based genetic algorithm: performance evaluation. *Applied Soft Computing* 1, pp. 285-299.

Hassan, F. A. (1978). Sediments in archaeology: Methods and implications for palaeoenvironmental and cultural analysis. *Journal of Field Archaeology* 5, pp. 197-213.

Kontogiorgos, D, Leontitsis, A. (2005). Micro-artefacts weight estimation by Genetic Algorithm minimisation, *Journal of Archaeological Science* 32, pp. 1275-1282.

Katayama, K., Sakamoto, H., and Narihisa, H. (2001). The efficiency of hybrid mutation genetic algorithm for the travelling salesman problem. *Mathematical and Computer Modelling* 31, pp. 197-203.

Kotsakis, K., and Halstead, P. (2004). Anaskafi sta Neolithika Paliambela Kolindrou. In *Arhaeologiko Ergo sti Makedonia kai sti Thraki* 16, pp. 407-415 (2002). Thessaloniki: Ministry of Culture, (in Greek).

Morgan, B. J. T. (2001). *Applied Stochastic Modelling*. Arnold.

Prügel-Bennet, A. (2004). When a genetic algorithm outperforms hill climbing. *Theoretical Computer Science*, to appear.

Rosen, A. M. (1989). Ancient Town and City Sites: A View from the Microscope. *American Antiquity* 54, pp. 564-578.

Schmitt, L. M. (2001). Theory of genetic algorithms. *Theoretical Computer Science* 259, pp. 1-61.

Sherwood, S. C., and Ousley, S. (1995). Quantifying microartifacts using a personal computer. *Geoarchaeology* 10, pp. 423-428.

Simms, S.R., and Heath, K.M. (1990). Site structure of the Orbit-Inn: An application of Ethnoarchaeology. *American Antiquity* 55, pp. 797-813.

Stein, J. K., and Telster, P. A. (1989). Size distributions of artifact classes: Combining Macro- and Micro-fractions. *Geoarchaeology*, 4 (1), pp. 1-30.

In: On Site Geoarchaeology on a Neolithic Tell Site ... ISBN 978-1-60741-366-0
Editor: D. Kontogiorgos © 2009 Nova Science Publishers, Inc.

Chapter 4

MICROARTELYZER: A SOFTWARE FOR MICROARTIFACTS' QUANTIFICATION

Alexandros Leontitsis[1,2], Dimitris Kontogiorgos[3]

[1] Department of Culture and New Technologies, University of Ioannina, 45110 – Panepistimioupoli, Ioannina, Greece [e-mail: me00743@cc.uoi.gr]

[2] Center for Research and Applications of Nonlinear Systems, University of Patras, 26500 – Rio, Patras, Greece

[3] Department of Archaeology and Prehistory, University of Sheffield, Northgate House, Sheffield, U.K.

4.1. INTRODUCTION

Several techniques for the quantification of microartifacts have been applied either by sorting and counting large number of samples (e.g., Simms and Heath, 1990); by estimating percentages of microartifact categories (e.g., Hassan, 1978; Rosen, 1989) or by point counting statistically representative samples (e.g., Stein and Telster, 1989; Sherwood and Ousley, 1995; Kontogiorgos and Leontitsis, 2005). The use of microartifacts in an archaeological analysis, however, is rarely undertaken because entails tentative microscopic identification, sorting and counting of different microartifact material classes (e.g., micro-pottery, micro-shell, micro-bone, etc.) until the researcher obtains a statistically representative sample (i.e., 95% error bounds). These disadvantages further required developing methods for facilitating microartifacts' quantification. Sherwood and Ousley (1995) report that by developing a computer program called MMCOUNT succeeded to reduce the intractable time factor in microartifacts' quantification making counting faster, while their program provides standard errors and calculates 95% error bound for each microartifact class entered in the program.

In this study we present a software for microartifacts' quantification, called *Microartelyzer*. It is developed by the Microsoft Visual C# Express edition, and it is running under Microsoft Windows. The heart of this program is the Particle Swarm Optimization (PSO hereafter) algorithm. It aims to reduce the point-counting time and effort providing an average estimation and the corresponding statistical errors of cultural (and non-cultural) sedimentary particles, when samples from archaeological deposits are analyzed for microartifacts using the point-counting method.

The structure of the study is the following. Section 2 reviews the concept and the technical details of the PSO, which is extensively used by Microartelyzer. Section 3 analyses the use of the Miroartelyzer. Finally, last section presents the conclusions of this study.

4.2. CONCEPT AND TECHNICAL DETAILS OF THE PSO

The PSO, proposed by Kennedy Eberhart (1995), is one of the most efficient optimization algorithms ever developed. There are applications on a broad band of science and engineering (e.g., Aruldoss Albert Victoire and Ebenezer Jeyakumar 2004, Kannan *et al.* 2004, Shi and Eberhart 1999).

The PSO algorithm was inspired by a flock of birds searching for food. Randomly distributed particles (in the case of optimization) move from one point to the other on the optimization space. Their movement is not fully deterministic, nor fully random. In between pure chance and pure determinism, step by step, through some kind of communication, the best solution is achieved.

Let us suppose that we have a search space of m dimensions. There are 2 equations that describe the above mentioned behavior.

$$x_{i,d} = x_{i-1,d} + v_{i,d} \tag{1}$$

$$v_{i,d} = wv_{i,d-1} + c_1 \text{rand}()(p_{i,d} - x_{i,d}) + c_2 \text{rand}()(P_d - x_{i,d}) \tag{2}$$

where x_i is an m-dimensional vector (particle), d=1,2,...,m, i is the number of iterations (also named *flights*) i=1,2,..., v is the m-dimensional vector of the velocity of each particle at the flight i, w is a constant between 0 and 1 which determines the autocorrelation in a particle's movement, p is the minimum found by the particle so far, and P is the minimum found by all the particles so far. In Microartelyzer we set w=0, because after experimentation we didn't observe any significant improvement to the results.

The matrix p and the vector P are responsible for the communication between particles, that we mentioned earlier. Finally, there are 2 constants (c_1 and c_2), which are usually given the value of 2, and a random number generator rand(), which gives uniformly distributed random numbers on [0,1].

The whole process is iterative. Each new iteration uses information by the previews iteration. It can be easily concluded that the more the particles are and the more the iterations one sets, the more efficient the algorithm becomes, but at the same time it becomes more time consuming.

4.3. MICROARTELYZER: DESCRIPTION OF THE PROGRAM, TECHNICAL DETAILS, APPLICATION AND RESULTS IN ARCHAEOLOGICAL MATERIAL

The Microartelyzer application developed having in mind the end-user who is not familiar with the technicalities described in section 2. We simplified the interface (Figure 4.1) as much as we could, requiring only a familiarity with Microsoft Excel.

4.3a. Data Preparation

The point counting data should be prepared like the data presented in Table 4.1. The first column should contain the sample weights. The order of the following columns is not important. For convenience we suggest the contexts to be on the rightmost part of the table.

When we are ready with the data preparation, we check the cells (including their names) and we copy them (Edit → Copy). In the case of table 1 we check the cells A1 to H29.

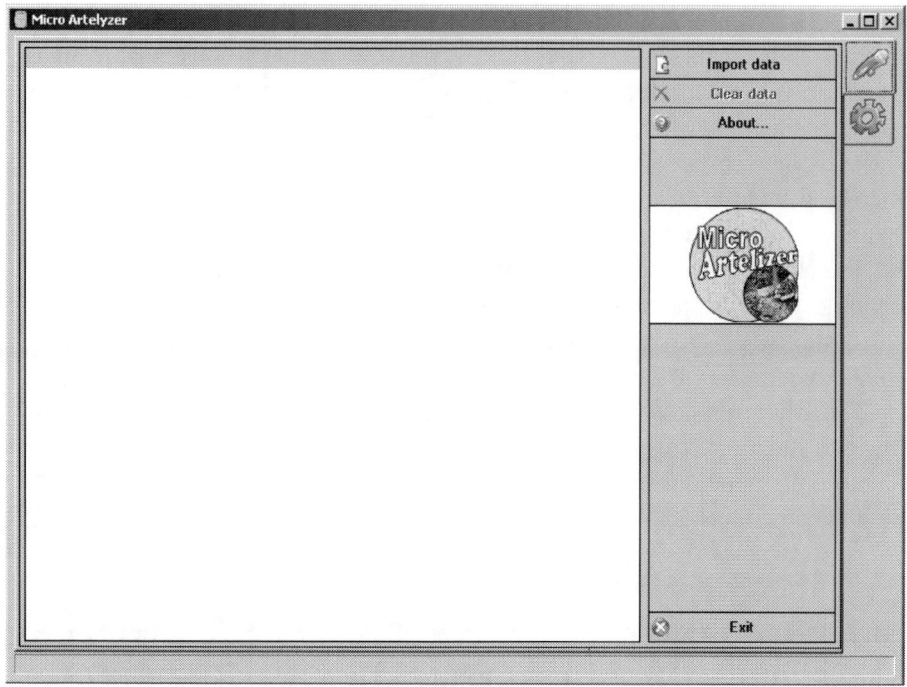

Figure 4.1. The simplified Graphical User Interface (GUI) of Microartelyzer.

Table 4.1. Data preparation using Microsoft Excel.

	A	B	C	D	E	F	G	H
1	Sample weight	Shell	Bone	Burnt Clay	Charcoal	Context 1	Context 2	Context 3
2	136.8	13	4	51	11	421	0	0
3	190.0	14	6	32	0	448	0	0
4	133.3	11	2	48	0	439	0	0
5	250.3	10	1	52	0	437	0	0
6	242.6	3	0	50	0	447	0	0
7	221.9	6	5	36	0	453	0	0
8	280.0	16	0	45	0	449	0	0
9	273.6	9	4	27	2	452	0	0
10	150.0	4	6	12	0	478	0	0
11	26.0	0	0	0	0	500	0	0
12	61.7	5	2	0	25	0	468	0
13	104.6	16	6	32	0	0	446	0

14	237.3	3	2	24	0	0	469	0
15	322.2	1	9	12	0	0	478	0
16	235.2	5	0	20	0	0	475	0
17	439.0	1	0	23	0	0	476	0
18	289.7	0	0	20	0	0	480	0
19	235.4	6	3	19	0	0	472	0
20	189.9	4	0	20	0	0	476	0
21	150.2	2	2	10	0	0	486	0
22	64.9	2	0	0	0	0	498	0
23	40.9	51	3	50	15	0	0	479
24	36.1	39	13	42	10	0	0	399
25	37.6	43	15	38	0	0	0	404
26	38.7	53	10	29	15	0	0	403
27	35.0	40	8	37	10	0	0	405
28	27.1	25	6	23	0	0	0	444
29	68.8	30	5	45	15	0	0	405

The sample weight should always be in the first column. The contexts are all together on the rightmost part of the table.

Usually the point counting takes into account 1000 particles for each sample. If there are more than 1000 particles counted, or less than 1000 particles counted, the program makes an internal adjustment. For example, in Table 4.1 all the rows add up to 500. The program multiplies all the particles by 2, in order to run the optimization and give the results as if the researcher had counted 1000 particles. We have to keep in mind that the accuracy for few data is relatively poor.

4.3b. Data Processing

To import the data into Microartelyzer, we click on the first link (*Import data*) (Figure 4.2). This way the data are pasted into Microartelyzer and become visible on the main screen. If we are not satisfied by the imported data, we may delete them (pressing the link *Clear data*), and go back to Microsoft Excel. The *About...* link gives the user the basic information about the program and the authors, and the *Exit* link terminates the program.

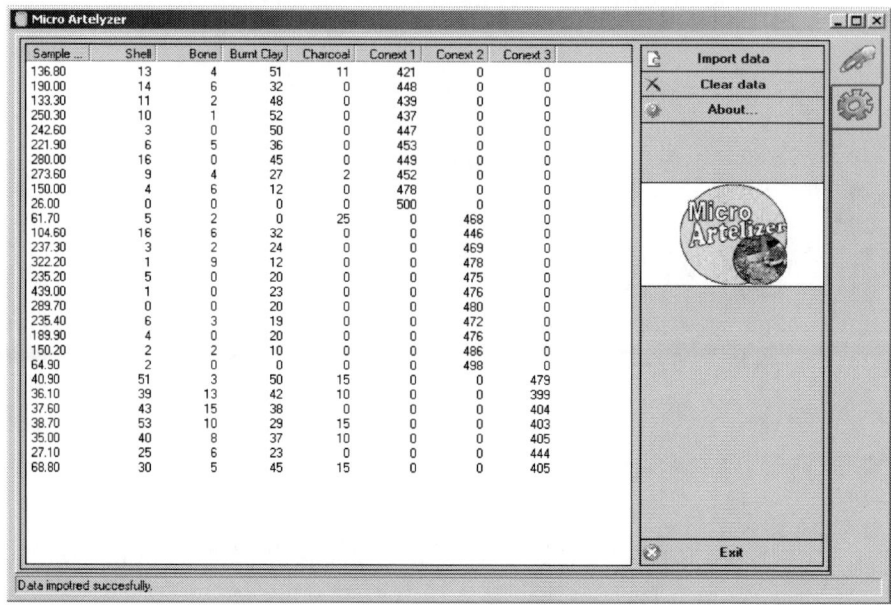

Figure 4.2. The screen that confirms that the data are imported.

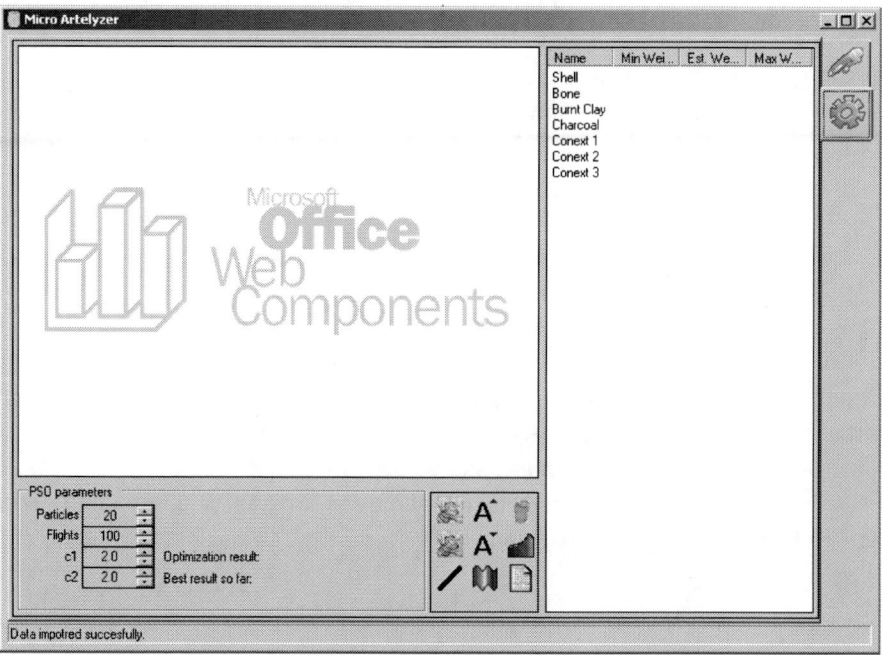

Figure 4.3 Pressing the tab with the gear icon, we go to the process screen.

Table 4.2. The explanations of the buttons' panel

Run optimization: Runs the optimization with the current parameter values.	**Increase font**: Increases the font size, indicating the current font size	**Export weights**: Copies to the clipboard the matrix on the right of the screen, with the final weight estimates for each microartefact and each context.
Actual vs Fitted: After the first optimization is finished, and thereafter, it compares the actual sample weigh to the estimated one (Figure 5)	**Decrease font**: Decreases the font size, indicating the current font size	**Export weights and their estimates**: Copies to the clipboard the Imported sample weight and the estimated sample weight.
Linear / Log plot: Changes the scale of the plot from linear to logarithmic and vice versa. It is useful to indentify small discrepancies.	**Color / Black White**: Changes the colors of the plot to black & white and vice versa	**Export Optimization**: Copies to the clipboard the iterations of the best optimization that was run so far.

The next step is to go to the process screen (pressing on the tab with the gear icon) (Figure 4.3). There is a panel of icons on the bottom on the screen. The explanations of these buttons are given on Table 4.2.

Once we go to the process screen, it is advisable to run the optimization a few times. After the second run, a red line will appear, indicating the best result so far (Figure 4.4). Every time a new minimum is reached, the particle weight estimates, on the table on the rightmost part of the screen, is updated.

The user has many options on data export, the most useful of which is the table with the particle weight estimates (Figure 4.5). The data export buttons are the rightmost buttons of the panel, and they all work in the same way. We press an export button, we switch to Microsoft Excel, and we paste the data (Edit → Paste).

A final note about the PSO parameters: Regarding c_1 and c_2 it is not advisable to change them, since the suggested values are suggested by the literature. The other two parameters (i.e., particles and flights) can be freely adjusted for experimentation.

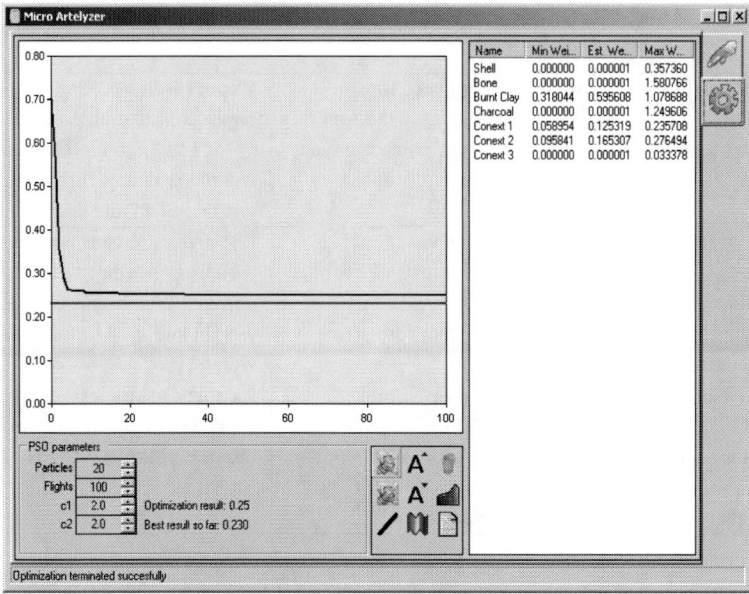

Figure 4.4. After few optimization runs, no result fell bellow the red line.

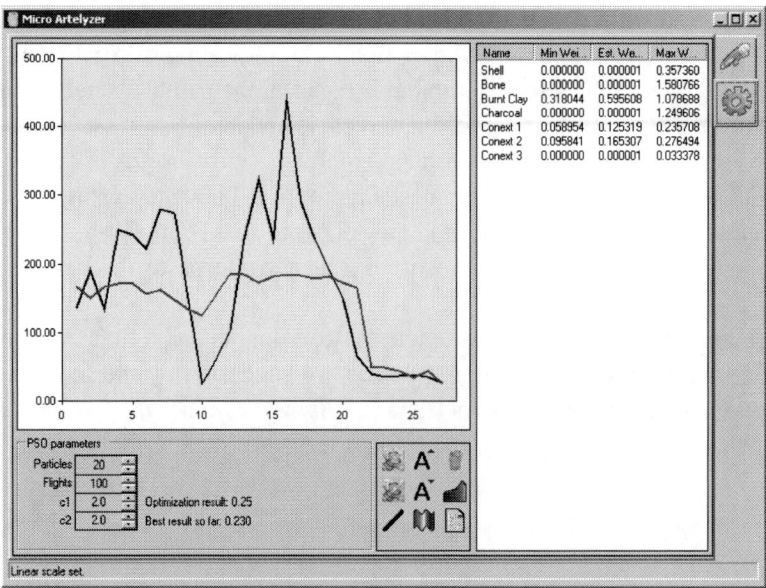

Figure4.5. The blue line corresponds to the actual sample weights. The read line corresponds to the sample weights that come from the estimated particle weights.

4.4. CONCLUSION

Microartifact analysis can assist in archaeological interpretation but is rarely undertaken due to time and effort cost in sample quantification. This study presented a Windows application called *Microartelyzer* that can free the researcher from additional counting time and effort but also provides a tool for determining the average weight of microartifacts', when samples from archaeological deposits are analyzed for microartifacts using the point-counting method. The software uses the Particle Swarm Optimization (PSO), one of the most promising optimization algorithms, and was applied experimentally on archaeological data.

The Microartelyzer software is available at http://leoaleq.ezeserv.com/.

REFERENCES

Hassan, F.A. (1978). Sediments in archaeology: Methods and implications for palaeoenvironmental and cultural analysis. *Journal of Field Archaeology* 5, pp.197-213.

Kennedy J., and Eberhart R. C. (1995). Particle swarm optimization. *Proc. IEEE International Conference on Neural Networks*, Perth, Australia, Nov. 27-Dec. 1, pp. 1942-1948.

Kontogiorgos, D, and Leontitsis, A. (2005). Micro-artefacts weight estimation by Genetic Algorithm minimisation, *Journal of Archaeological Science* 32, pp. 1275-1282.

Aruldoss Albert Victoire T., Ebenezer Jeyakumar A. (2004). Hybrid PSO–SQP for economic dispatch with valve-point effect. *Electric Power Systems Research* 71(1), pp. 51-59.

Kannan S., Mary Raja Slochanal S., Subbaraj P., Narayana Prasad Padhy. (2004). Application of particle swarm optimization technique and its variants to generation expansion planning problem. *Electric Power Systems Research* 70(3), pp. 203-210.

Rosen, A.M. (1989). Ancient Town and City Sites: A View from the Microscope. *American Antiquity* 54, pp. 564-578.

Sherwood, S.C., and Ousley, S. (1995). Quantifying microartifacts using a personal computer. *Geoarchaeology* 10, pp.423-428.

Shi Y., and Eberhart R. C. (1999). Empirical study of particle swarm optimization. *Congress on Evolutionary Computation*, July 6-9 1999, pp. 1945-1950.

Simms, S.R., and Heath, K.M. (1990). Site structure of the Orbit-Inn: an application of Ethnoarchaeology. American Antiquity 55, pp.797-813.

Stein, J.K., and Telster, P.A. (1989). Size Distributions of Artifact Classes: Combining Macro- and Micro-Fractions. *Geoarchaeology*, 4(1), pp. 1-30.

In: On Site Geoarchaeology on a Neolithic Tell Site ... ISBN 978-1-60741-366-0
Editor: D. Kontogiorgos   © 2009 Nova Science Publishers, Inc.

Chapter 5

FROM TELLS TO EXTENDED SETTLEMENTS: ON-SITE GEOARCHAEOLOGY AND CULTURAL FORMATION PROCESSES AT THE EXTENDED NEOLITHIC SETTLEMENT AT KORINOS (NORTHERN GREECE)

Manthos Besios[1], Fotini Adaktylou[1], Dimitris Kontogiorgos[2]

[1] KZ' Ephoreia of Prehistoric and Classical Antiquities 32, Parmenionos Str. 601 00 Katerini, GR.

[2] Department of Archaeology and Prehistory, University of Sheffield, Northgate House, Sheffield, U.K.

5.1. INTRODUCTION: SITE-TYPE VARIABILITY, SITE FORMATION PROCESSES AND GEOARCHAEOLOGY

The conventional model of Greek Neolithic society has been dominated by the results of excavations in Thessaly in the early 20th century by Tsountas (1908) and Wace and Thompson (1912). The resulting picture of a society of rectangular 'family houses' of mud brick, grouped into compact and long-lived villages that ultimately developed into 'tell' mounds, passed into many syntheses of both Aegean (e.g., Theochares, 1973) and European (e.g., Childe, 1957) prehistory.

In the past one or two decades, this picture has been challenged in many ways. It has been argued that rectangular mud brick houses represent an ideological statement as much as domestic shelter (e.g., Hodder, 1990; Kotsakis, 1999), that early houses were flimsy and early settlements short-lived or seasonal (Whittle, 1997) and that architectural definition of 'household' units took place gradually over several millennia (Halstead, 1995). It has also been recognised that Neolithic tell settlements developed alongside 'flat-extended' sites, especially in northern Greece. The latter type of site is more easily buried or eroded and less easily detected by extensive surface reconnaissance (e.g., Krahtopoulou, 2001) and may also have marked a less prominent 'place' in the Neolithic cultural landscape (Chapman, 1989; Kotsakis, 1999).

The identification of the extended settlements led archaeologists to fruitful discussions of the social and economic organization, as well as the ideological meaning of this variability in site type during the Greek Neolithic. The outcome of these discussions was to show that the difference between tell and extended sites can be understood on multiple levels; it was argued (Kotsakis, 1999: 69-74) that the picture of considerable continuity, spatial and temporal, that is often represented in the archaeological deposits of tell sites clearly contrasts with the spatial and temporal discontinuity of the deposits at the extended settlements. The eventual domination of tell settlements in the Greek Neolithic, the construction of monumental houses and hence the process of tell formation may be related to household definition and intra-household competition (Kotsakis, 1999; Halstead, 1999).

Identifying and explaining site formation processes is an important first step towards developing justifiable inferences about past behaviour and past societies. A sound understanding of the archaeological record and how it was formed is necessary in order to effectively interpret the data that is recovered. The importance of the study of formation processes remains critical for developing better inferences, but also to help illuminate the organisation of and change in behavioural systems of the past (e.g., Schiffer 1983; Goldberg *et al.* 1993; McGuire 1995; Reid 1995; Tani 1995; LaMotta and Schiffer, 2001). The unit of analysis appropriate for identifying formation processes is the deposit. Schiffer (1987:265) defines a deposit as "a three-dimensional segment of a site (or other area of analytical interest) that is distinguished in the field on the basis of observable changes in sediments and artefacts."

Although it is common practice, viewing the deposit as a single discrete depositional event or process has its problems, as a single depositional process can give rise to materials in different deposits, and conversely, a single deposit can contain the products of many different depositional processes (Schiffer 1987:266). Despite these issues, a variety of formation processes can be identified through the traces of evidence they leave. Geoarchaeology can add substantially to the study of site formation processes, and its current applications are an attempt to provide answers about how the archaeological record was formed and how it was preserved. This type of understanding clarifies what in the archaeological record is a reflection of past culture and what is a reflection of formation processes.

Finally, of equal importance is taking individual contexts as the unit of analysis because they comprise many of the components (e.g., bones, seeds, pottery, etc.) that create patterning and variation in the archaeological record. A recent geoarchaeological study of pits and ditches from a Neolithic Tell site in Greece suggested that the variability in their sedimentary and artefactual characteristics might be attributed to diverse attributes of past human activities, and that the structure of debris in these contexts was the result of cultural formation processes. Also, it was possible to discern between primary and secondary formation processes in the ditches (Kontogiorgos, 2008).

On-site geoarchaeological work at the extended Neolithic site at Korinos (Pieria region, Northern Greece) was thus initiated with the aims of investigating the formation processes occurring in one of the deepest archaeological contexts, a deep pit, detected on that site and interpreted as a semi-subterranean dwelling. The definite interpretation of the initial function of this kind of architectural structures is relatively difficult because their fills comprise abundance of cultural debris that could be attributed either to habitation (e.g., Bogdanovic, 1988) or refuse disposal (e.g., Gimbutas, 1991), while in many cases these pits, dug into the natural bedrock of a site, were used for the extraction of primary building material (Gimbutas et al, 1989). In other cases, there is a serious difficulty to discern the function of a pit, as in the case of Starcevo-Cris-Koros culture area (e.g., Jongsma, 1997).

The following geoarchaeological example, although not exhaustive, is hoped to provide some additional information regarding the cultural (and natural) formation processes occurring in these types of deposits while it will test the possibility to discern between primary and secondary formation processes. The structure of this study is the following: section 5.2 and 5.3 describe the excavated site and the sampling procedure/laboratory methods, respectively; section 5.4 presents the results of the geoarchaeological analysis while section 5.5 offers the conclusions of this study.

5.2. THE SITE: LOCATION AND EXCAVATED CONTEXTS

The archaeological site is located to the west of the modern village of Korinos (Northern Pieria region, Northern Greece) (Figure 5.1) and excavation during three seasons (2002-2004) has confirmed the presence of significant constructions, including 86 Early Neolithic storage pits and pit-houses (semi-subterranean dwellings) (Figure 5.2), and a total of 28 postholes that were interpreted as traces of a surface house with rectangular bases. Finally, rich pottery finds date the extended settlement to the Early Neolithic period (Besios and Adaktylou, 2004). One of the deepest Early Neolithic pits, pit 24, ranging in size from a maximum length of 3,70m (N-S) to 3,35m wide (E-W), was selected for geoarchaeological investigation. No internal structures were detected inside pit 24 while a circular pit, pit 55, cut by pit 24, may possible interpreted as an entrance ramp to pit 24, thus suggesting a possible interpretation of pit 24, as a pit-house. Early Neolithic pit 24 (thereafter: EN pit 24) was filled with cultural debris mainly animal bones, stone and bone tools, shells, and burnt clay fragments.

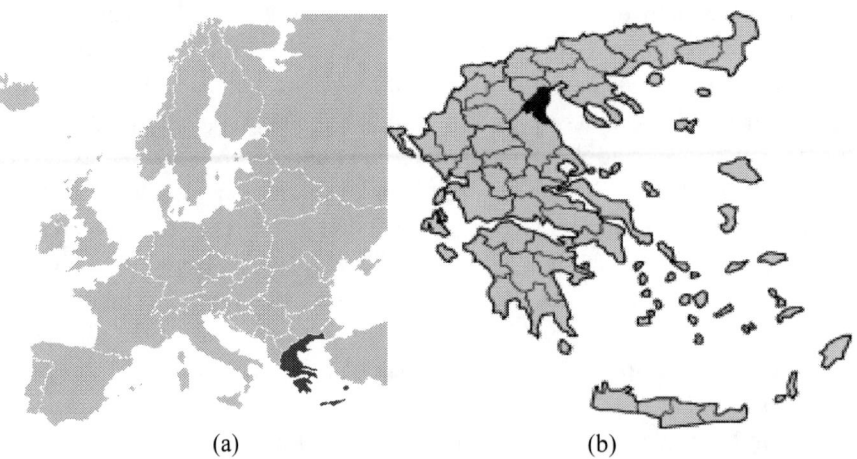

Figure 5.1. a) Map of Europe, showing the location of Greece, b) Map of Greece, showing the location of the Pieria region.

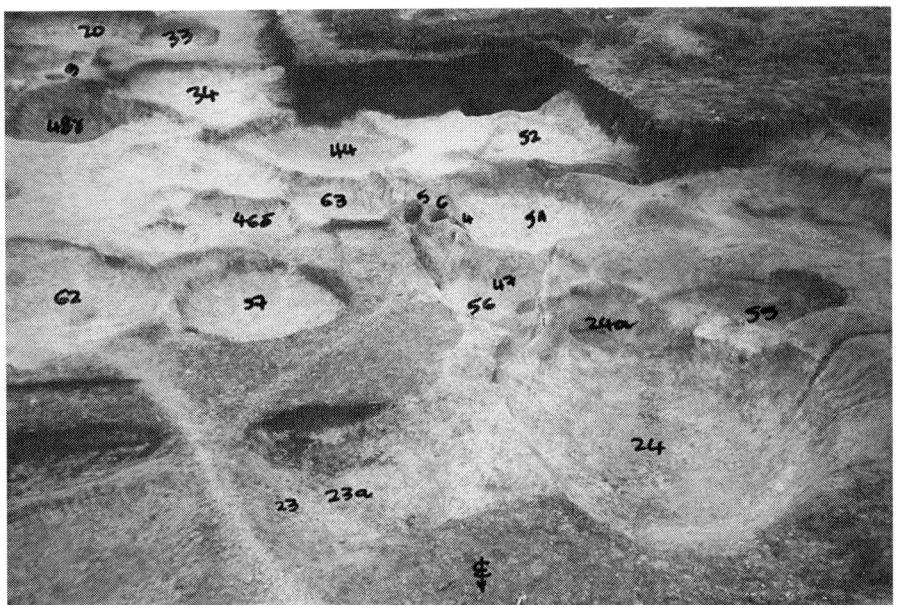

Figure 5.2. View (N-S) of the excavated site and EN pit 24 (Source: 'Korinos excavation archive', KZ' Ephoreia of Prehistoric and Classical Antiquities, Greek Ministry of Culture).

5.3. SAMPLING AND METHODOLOGY

Macroscopic examination of the EN pit 24 clearly defined five basic stratigraphic units on the profile of the excavated context: a silty layer (10YR 5/6) at the very bottom of the feature and a coarser, gravelly, layer sitting on top of it; two thick, silty, brownish layers (10YR 4/4 (lower layer)-10YR 4/2 (upper layer)), rich in cultural materials, macroscopically divided on the basis of colour, were discerned above the gravelly layer; and a 'top-soil' layer (10YR 3/1) (Figure 5.3). It should be noted, however, that sampling was not proportional to the thickness and variability of the fill layers since macroscopic examination made clear that the layers were too complicated to accommodate such a sampling procedure, while more than one sample was taken from the clearly distinguished basic stratigraphic units.

Figure 5.3. EN pit 24-Field stratigraphy (View N-S) (Source: (Source: 'Korinos excavation archive', KZ' Ephoreia of Prehistoric and Classical Antiquities, Greek Ministry of Culture).

Table 5.1. Gravel, sand, silt, clay weight percentages for EN pit 24

gravel	sand	silt	clay
3.27	15.63	55.30	25.80
4.38	12.56	53.53	29.53
6.25	13.18	55.94	24.63
6.73	12.66	55.28	25.33
7.67	11.41	53.13	27.79
6.95	11.74	52.66	28.65
3.61	10.97	56.38	29.04
5.75	10.66	53.61	29.98
3.12	11.61	56.36	28.91
3.37	11.43	60.45	24.75
8.9	10.99	59.04	21.07
9	9.82	59.58	21.60
2.66	9.94	52.87	34.53

2.62	9.28	52.41	35.69
3.16	12.34	56.77	27.73
5.29	13.92	54.22	26.57
7.39	15.47	54.22	22.92
7.64	13.86	49.79	28.71
9.87	15.82	51.90	22.41
16.52	17.43	45.21	20.84
4.32	24.28	51.68	19.72
7.61	15.29	49.49	27.61
24.2	22.76	40.09	12.95

A total of 23 sediment samples, with an average weight of ca 1500g, were collected in columns mostly at 5cm vertical intervals on the profile of the context and were labeled according to depth. Two methods of analysis were applied in the sediment samples: particle size analysis for texture determination, and microartifact analysis for the cultural sedimentary particles smaller than 2mm in diameter (e.g., bone, shells, etc.). For the determination of particle size, hydrometer analysis (ASTM, 1961) was used for silt and clay, and sieving for the coarse particles (gravel and sand) (e.g., Folk, 1980) (Table 5.1). The procedure for determining the proportions of microartifact compositional types follows that described by Stein and Telster (1989: 10-11). One of the state-of-the-art minimization methodologies, the so-called Genetic Algorithms was applied for microartifact density determination (Kontogiorgos and Leontitsis, 2005; Chapter 3, this volume) (Tables 5.2-5.4).

Table 5.2. Microartifact data-base for EN pit 24

Total weight	Sample weight	Shell	Bone	Burnt Clay	Charcoal	Sediment
1465.6	23.9	0	29	60	5	407
1440.3	38.1	17	25	99	0	359
1839	48.7	25	19	25	0	431
1113	29.2	25	19	25	0	431
847.3	33.5	16	7	20	0	457
864	18	10	10	40	0	440
969.5	20.9	10	24	50	0	416
1043.1	17.5	10	24	50	0	416
1439.9	26.1	11	60	40	0	389

Table 5.2. (Congress)

Total weight	Sample weight	Shell	Bone	Burnt Clay	Charcoal	Sediment
1336.2	27.8	7	20	20	0	453
2022.9	47.5	4	22	35	0	439
1776	40.1	7	48	25	0	420
1500.1	14	9	12	27	2	450
1483.5	16.6	0	7	14	0	479
1331.3	20.4	3	3	10	0	484
1892.6	41.7	6	4	14	3	473
2030	64.6	6	11	14	27	442
1377.1	46.9	5	12	20	47	416
2127.1	98.6	5	12	20	47	416
1223.1	35.1	2	46	31	0	421
1505.2	33.6	0	35	60	0	405
1640.5	48	6	39	21	0	434
1157.2	94.1	5	0	23	0	472

Table 5.3. Microartifact weights per cultural particle and confidence intervals for EN pit 24

	95% Lower Bound	Estimation	95% Upper Bound
Shell	0.0000	0.0285	0.7019
Bone	0.0000	0.0002	0.5004
BurntClay	0.0000	0.0002	0.2354

Table 5.4. Microartifact density for EN pit 24

Shell	Bone	Burnt Clay
0.00%	0.05%	0.08%
2.55%	0.03%	0.08%
2.93%	0.02%	0.02%
4.89%	0.03%	0.03%
2.73%	0.01%	0.02%
3.17%	0.02%	0.07%
2.73%	0.05%	0.07%

3.26%	0.06%	0.09%
2.41%	0.10%	0.05%
1.44%	0.03%	0.02%
0.48%	0.02%	0.02%
1.00%	0.05%	0.02%
3.67%	0.04%	0.06%
0.00%	0.02%	0.03%
0.84%	0.01%	0.02%
0.82%	0.00%	0.01%
0.53%	0.01%	0.01%
0.61%	0.01%	0.01%
0.29%	0.01%	0.01%
0.33%	0.06%	0.03%
0.00%	0.04%	0.05%
0.71%	0.03%	0.01%
0.30%	0.00%	0.01%

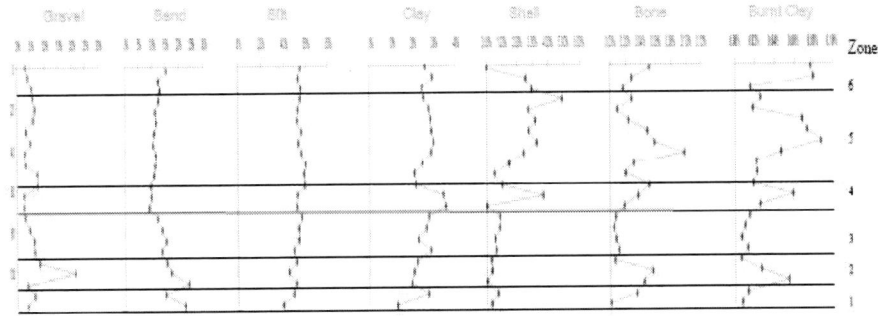

Figure 5.4. Particle size and microartifacts for EN pit 24.

5.4. EN PIT 24: PIT-HOUSE OR JUST A 'PIT'? (FIGURE 5.4)

Particle size analysis and microartifacts defined the stratigraphy of the EN pit 24. The lower part of the pit (Zone 1) comprises rising clay and microartifact values possibly compatible with slow sedimentation and increased cultural material accumulation. Rising values in bone and burnt clay suggest relatively fast accumulation of these materials in the lower part of Zone 2, until this accumulation was interrupted by rapid, coarse sediment input (gravel peak) in the middle/upper part of Zone 2. In Zone 3, texture and cultural micro-materials exhibit stable/modest values suggesting moderate cultural deposition. Zone 4 is characterized by acceleration in clay input and co-variable values for shell and burnt clay. These characteristics suggest fine, slow, sedimentation that might have been responsible for the co-variation in shell and burnt clay and the upwards rising values of bone. Zone 5, clearly contrasts with Zone 4, exhibiting variable values for all cultural micro-materials and almost stable texture suggesting variability in cultural material deposition. Finally, the upper part of the pit (Zone 6) exhibits relatively stable texture and variability in microartifact density trend.

Thus, the analysis of sediments and microartifacts clearly demonstrates a lower fill (i.e., Zone 1 and lower part of Zone 2) attributable possibly to cultural deposition, that was interrupted by fast, coarse (gravelly) sediment input (i.e., upper part of Zone 2) which may demonstrate temporary abandonment in cultural use of the feature, and, as result, possible fall of its sides (with the later explanation being compatible with the macroscopic examination of the profile). After this phase of possible cultural disuse of the pit, cultural deposition continues slowly in Zone 3. This low rate of cultural material accumulation might have been temporarily interrupted by a phase of fine sediment acceleration in Zone 4, plausibly resulting from exposure of this part of the pit to rainfall (see also Kontogiorgos, 2008 for interpretation of clayish layers in pits).

Cultural infilling of the pit continues increased and variable again in Zone 5, possibly demonstrating variability in cultural activities ending up in the pit. Finally, the formation of upper part of the pit (i.e., Zone 6) may be attributed to variability in cultural material accumulation but since it is just underneath the modern, agricultural top-soil, the sedimentary and microartifactual picture may have been seriously mixed up and transformed by modern agricultural practices.

The geoarchaeological analysis provide some grounds to support the existence of an initial, primary fill in EN pit 24 (i.e., Zone 1-lower part of Zone 2) resulting possibly from primary cultural formation processes. Whether these primary cultural processes, at the bottom of EN pit 24, represent habitation activities, occurring inside the pit, or the first phases of cultural debris/refuse deposition, is unclear. Simply with the present data in hand is difficult to assess whether EN pit 24 was a pit-house or just a multifunctional pit, receiving the end up results of different human activities, during the first stages of its use life. The later interpretation, as a multifunctional 'tank', fits better with the characteristics of the upper fill of EN pit 24, further indicating secondary cultural formation processes. In short, the analysis of sediments and microartifacts in EN pit 24, although strongly depicts primary and secondary cultural formation processes as responsible for its infill formation (at least for its largest part), however, do not pinpoint to a specific use of the pit as a pit-house.

5.5. CONCLUSIONS

The geoarchaeological study in one of the many EN pits (i.e., EN pit 24) detected at the extended EN settlement at Korinos (Pieria region, Northern Greece) and interpreted either as pit-houses and/or storage pits, has succeeded in clarifying the stratigraphy of the feature, demonstrating, also, that cultural formation processes contributed significantly to its formation. An attempt to investigate whether EN pit 24 was as a pit-house, revealed the necessity to further investigate geoarchaeologicaly this type of deposits in order to establish strong patterning before any further interpretation. In any case, this attempt is the first step towards this goal and call for the awareness with which to view and interpret one of the commonest archaeological features that is, pits.

REFERENCES

ASTM (American Society for Testing Materials) (1961). Tentative method for grain-size analysis in soils. In *The 1961 Book of ASTM Standards*, pt.4, pp.1272-1283.

Besios, M., and Adaktylou, F. (2004). Neolithikos Ikismos sta 'Ravenia' Korinou. *Paper presented at Archaeologikon Ergon sti Makedonia and sti Thraki Conference 2004.* (In Greek).

Bogdanovic, M. (1988). Architectural and structural features at Divostin. In Mc Pherron, A., and Srejovic, D. (eds.), *Divostin and the Neolithic of Central Serbia*. University of Pittsburgh, Department of Anthropology, Ethnology Monographs, 10, pp.35-42.

Chapman, J. (1989). The early Balkan Village, In S. Bökönyi (ed.) *Neolithic of Southeastern Europe and its Near Eastern Connections*, pp. 33-53. Budapest: Varia Archaeologica Hungarica.

Childe, V.G. (1957). *The Dawn of European Civilisation*. London/New York: Rotledge and Keagan/1957.

Folk, R.L. (1980). *Petrology of Sedimentary Rocks*. Austin: Hemphil.

Gimbutas, M., Winn, S., and Shimabuku, D. (1989). Achilleion: A neolithic Settlement in Thessaly, Greece, 6400-5600 B.C. *Monumenta Archaeologica 14*. Los Angeles: University of California.

Gimbutas, M. (1991). *The Civilization of Goddess: The World of Old Europe*. New York. Harper.

Goldberg, P., D. T. Nash and M. D. Petraglia. (1993). *Formation Processes in Archaeological Context*. Madison, Wisconsin: Prehistory Press.

Halstead, P. (1995). From sharing to hoarding: the Neolithic foundations of Aegean bronze age society? In Laffineur, R. and Niemeier, W.-D. (eds.), *Politeia: Society and State in the Aegean Bronze Age* (Aegaeum 12), pp. 11-20. Liege: University of Liege.

Halstead, P. (1999). Neighbours from Hell? The Household in Neolithic Greece. In Halstead, P. (ed.), *Neolithic Society in Greece*, pp. 77-95. Sheffield Studies in Aegean Archaeology 2, Sheffield Academic Press.

Hodder, I., (1990). *The Domestication of Europe*. Oxford. Basil Blackwell.

Jongsma, T. (1997). Distinguishing pit from pit-houses through daub analysis: The nature and Location of Early Neolithic Starcevo-Cris Houses AT Foeni-Salas, Romania. *M.A thesis,* UMI, .Department of Anthropology, University of Manitoba.

Kontogiorgos, D, and Leontitsis, A. (2005). Micro-artefacts weight estimation by Genetic Algorithm minimisation, *Journal of Archaeological Science* 32, pp. 1275-1282.

Kontogiorgos, D. (2008). Geoarchaeological and Microartifact Analysis of Archaeological Sediments. *A Case study From a Neolithic Tell Site in Greece.* Nova Science Publishers, Inc.

Kotsakis, K. (1999). What Tells Can Tell: Social Space and Settlement in the Greek Neolithic. In Halstead, P. (ed.), *Neolithic Society in Greece*, pp. 66-76. Sheffield Studies in Aegean Archaeology 2, Sheffield Academic Press.

Krahtopoulou, A. (2001). *Late Quaternary Alluvial History of Northern Pieria, Macedonia, Greece.* Unpublished Ph.D dissertation, University of Sheffield, 2001.

LaMotta, V., M. and M. B. Schiffer. (2001). Behavioural Archaeology: Toward a New Synthesis. In I. Hodder (ed.) *Archaeological Theory Today*, pp. 14-64. Cambridge: Polity Press.

McGuire, R. H. (1995). Behavioural Archaeology: Reflections of a Prodigal Son. In J. M. Skibo, W. H.Walker and A. E. Nielsen (eds.) *Expanding Archaeology*, pp. 162-177. Salt Lake City: University of Utah Press.

Reid, J. J. (1995). Four Strategies after Twenty Years: A Return to Basics. In J. M. Skibo, W. H. Walker and A. E. Nielsen (eds.) *Expanding Archaeology*, pp. 15-21. Salt Lake City: University of Utah Press.

Schiffer, M.B. (1983). Toward the Identification of Formation Processes. *American Antiquity* 48, pp. 675-706.

Stein, J.K., and Telster, P.A. (1989). Size Distributions of Artifact Classes: Combining Macro- and Micro-Fractions. *Geoarchaeology*, 4(1), pp. 1-30.

Schiffer, M.B. (1987). *Formation Processes of the Archaeological Record*. University of New Mexico Press, Albuquerque.

Tani, M. (1995). Beyond the Identification of Formation Processes: Behavioural Inference Based on Traces Left by Cultural Formation Processes. *Journal of Archaeological Method and Theory* 2(3), pp. 231-252.

Theocharis, D.R. (1973). *Neolithic Greece*. Athens: National Bank of Greece.

Tsountas, Kh. (1908). *Ai Proistorikai Akropolis Diminiou kai Sesklou*. Athens. Arkhaiologiki Etairia.

Wace, A.J.B., and Thopson, M.S. (1912). *Prehistoric Thessaly*. Cambridge: CambridgeUniversity Press.

Whittle, A. (1997). Moving on and moving around: Neolithic settlement mobility. In Topping, P. (ed.), *Neolithic Landscapes* (Oxbow Monograph 86), pp.15-22. Oxford: Oxbow Books.

In: On Site Geoarchaeology on a Neolithic Tell Site ... ISBN 978-1-60741-366-0
Editor: D. Kontogiorgos © 2009 Nova Science Publishers, Inc.

Chapter 6

FROM NEOLITHIC TO HELLENISTIC. A GEOARCHAEOLOGICAL APPROACH TO THE BURIAL OF A HELLENISTIC THEATRE: THE EVIDENCE FROM PARTICLE SIZE ANALYSIS AND MICROARTIFACTS

Dimitris Kontogiorgos[1], Kaliopi Preka-Alexandri[2]

[1] Department of Archaeology, University of Sheffield, Northgate House, Sheffield, U.K.

[2] Ephoreia of Underwater Antiquities, 30 Kallisperi Str., 117 42, Athens, GR.

6.1. INTRODUCTION

The theatre of the Hellenistic period (ca. 330 B.C-167 B.C) is located outside the city walls of the Hellenistic city of Gitana (Thesprotia region-Epirus-NW Greece) (Figure 6.1). The systematic excavation during six seasons (1996-1997 and 2005-2008) has brought into light the theatre below a thick (ca. 1.25m-1.50m) colluvial deposit. The source of the colluvial deposit was thought to be sediments and cultural materials eroded from the abandoned Hellenistic city of Gitana, once expanded on top of the theatre (Figure 6.2). The geoarchaeological study of this deposit was thus initiated in order to further explore the processes responsible for its formation. The first section (6.2) briefly presents field and laboratory methods.

The second (6.3) and third (6.4) sections offer the results and conclusions of this study, respectively.

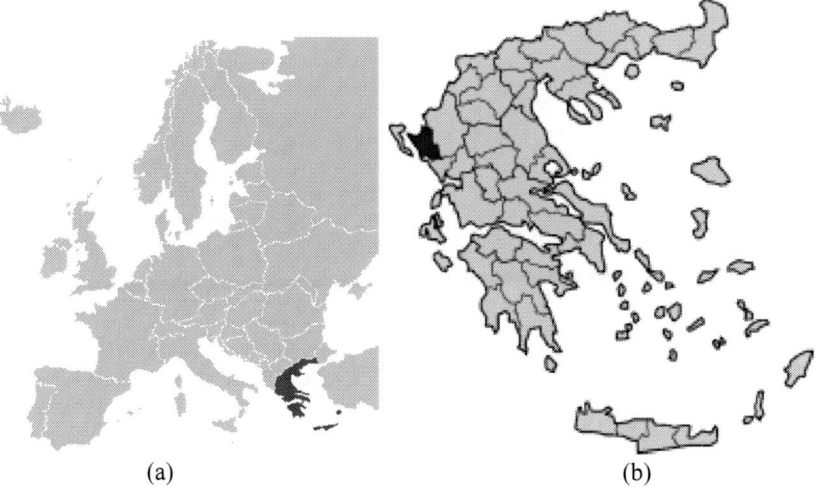

(a) (b)

Figure 6.1. (a) Map of Europe, showing the location of Greece, (b) Map of Greece, showing the location of Thesprotia region.

Source: 'Gitani excavation archive'- Preka-Alexandri,K.

Figure 6.2. The Hellenistic Theatre of Gitana.

6.2. FIELD AND LABORATORY METHODS

A total of forty four (44) sediment samples weighing ca 1000g each, were collected from five (5) columns, mostly at 10cm vertical intervals, providing good coverage across the exposed stratigraphy of the deposit (Figures 6.3-6.5), and were labeled according to depth. Macroscopic examination clearly defined two basic stratigraphic units on the exposed stratigraphy: the colluvial deposit and a 'top-soil' layer. Two methods of analysis were applied in the sediment samples: particle size analysis for texture determination (Tables 6.1-6.5) (e.g., ASTM, 1961; Folk, 1980), and microartifact analysis for the cultural sedimentary particles smaller than 2mm in diameter (e.g., bone, shells, etc.) (see details in Chapter 3, this volume).

Source: 'Gitani excavation archive'- Preka-Alexandri, K.

Figure 6.3. The colluvial deposit.

Source: 'Gitani excavation archive'- Preka-Alexandri, K.

Figure 6.4. Sampling and location of the profiles on the colluvial deposit.

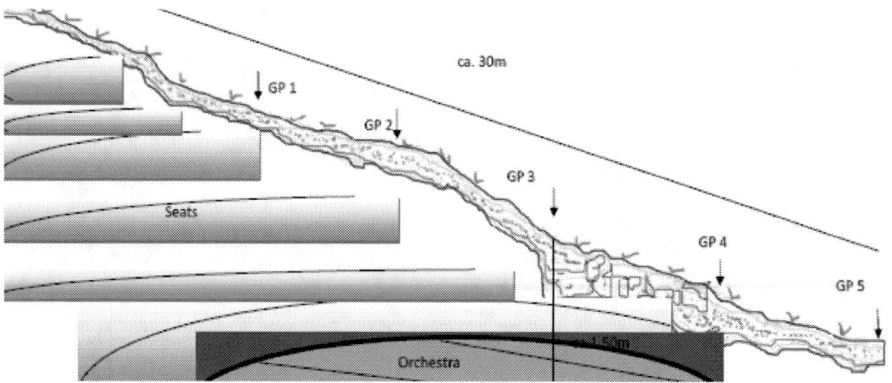

Figure 6.5. Stratigraphy of the colluvial deposit and location of the profiles (Schematic-not in scale).

Table 6.1. Gravel, sand, silt, and clay weight percentages for GP1

gravel	sand	silt	clay
22,53	28,18	38,91	10,37
32,89	23,95	33,19	9,96
17,6	21,72	44,16	16,51
10,04	22,48	48,84	18,63
8,85	21	51,14	19,00

Table 6.2. Gravel, sand, silt, and clay weight percentages for GP2

gravel	sand	silt	clay
31,77	31,54	32,70	3,98
43,41	25,26	26,76	4,56
49,8	20,65	24,23	5,31
27,69	24,47	35,79	12,04
26	23,04	37,60	13,35
15,88	29,21	39,78	15,12
14,21	24,57	45,92	15,29
20,37	19,67	46,33	13,62
26	21,13	40,23	12,63
29,89	18,18	40,38	11,54

Table 6.3. Gravel, sand, silt, and clay weight percentages for GP3

gravel	sand	silt	clay
14,52	25,76	44,82	14,89
16,01	24,87	43,82	15,29
20,36	24,22	41,13	14,28
13,95	30,94	40,84	14,26
21,26	26,98	38,08	13,67
31,18	22,92	37,77	8,12
32,29	20,88	36,21	10,61
17,41	24,95	44,84	12,79
18,98	25,13	42,31	13,57
7,14	29,28	49,10	14,47
9,44	26,15	49,72	14,68

Table 6.4. Gravel, sand, silt, and clay weight percentages for GP4

gravel	sand	silt	clay
2,98	30,06	48,22	18,73
3,59	28,8	44,65	22,95
6,22	30,37	43,47	19,93
0,89	23,78	52,94	22,38
4,46	16,8	55,37	23,36
6,09	19,75	52,40	21,75
14,2	22,8	46,60	16,39
19,72	24,68	40,65	14,94
12,99	25,26	45,21	16,53

Table 6.5. Gravel, sand, silt, and clay weight percentages for GP5

gravel	sand	silt	clay
2,71	21,85	58,79	16,64
11,34	22,41	47,51	18,73
5,8	26,73	50,85	16,61
10,94	24,22	48,53	16,30
11,8	22,1	49,91	16,18
13,67	22,48	46,22	17,62
25,13	16,88	44,74	13,24
23,19	16,81	45,68	14,31
19,65	19	44,38	16,96

6.3. RESULTS AND INTERPRETATION

The analysis of sediments and microartifacts indicate that the colluvial deposit consists, mainly, of fine textured (abundance of silt and clay particles) sediment and variable quantities of cultural micro-materials while there is also evidence of acceleration in input of coarser (gravel sized) sedimentary particles (i.e., GP3). More clearly, the lower part of the deposit (i.e., Zone 1 in GP 5, GP 4, and GP 3) exhibits rising upwards trend in cultural micro-materials and relatively stable input of finer sedimentary particles (silt and clay). Despite the uprising gravel values, fine sediment dominates Zone 1 (i.e., in GP 5, GP 4, and GP 3) suggesting relatively slow sedimentation that might have been responsible for microartifact generation.

Fine sediments, indicative possibly for slow sedimentation and variable quantities of cultural micro-materials dominate Zone 2 in GP 5 and GP4, while gravel sized sediment drops, rising again slightly in the uppermost part of this zone. Zone 2 in GP 3, exhibits upwards increases in gravel sized sediment, but the zone is again dominated by fine sedimentary particles and displays upwards increases in microartifacts. The gravel peak in Zone 2 in GP3 may be over-represents the similar high values of gravel at the lowest part of the sequence (i.e., the boundary between Zone 1 and Zone 2 in GP 5 and GP 4) due to the higher position of this profile in the sequence. Zone 3, in GP 3, exhibits similar sedimentary (i.e., domination of finer sedimentary particles; drop and then rise in gravel) and, to a lesser extent, microartifact characteristics (i.e., less variable and abundant) with Zone 2, in GP 5 and GP 4. Thus, the sedimentary and microartifactual characteristics of the lower part of the colluvial sequence, represented by GP5, GP4, and GP 3, are more compatible with slow sedimentation possibly of variable intensity, depicted in the variability of microartifacts (Figures 6.6-6.8).

At the upper part of the colluvial sequence, GP 2 and GP 1, both, exhibit broadly similar sedimentary and microartifact characteristics; Zone 1 comprises abundance of silt and clay and rising values in microartifacts, while in Zone 2 there is an increase in gravel, at the expense of silt and clay. Microartifacts drop slightly in Zone 2, in GP 2, but continue upwards in Zone 2, in GP 1. These characteristics imply slow sedimentation triggering possibly microartifact generation in Zone 1 (i.e., in GP2 and GP 1), while there is an indication for a gentle acceleration in coarse sediment input in Zone 2 (i.e., in GP2 and GP 1) (Figures 6.9-6.10). Finally, the darker colour of the 'top-soil' layer (i.e., Zone 3 in GP1, GP 2, GP 4 and GP 5, and Zone 4 in GP 3), comprising modest amounts of cultural micro-materials, is consistent with the development of an A-soil Horizon (Birkeland, 1999:11).

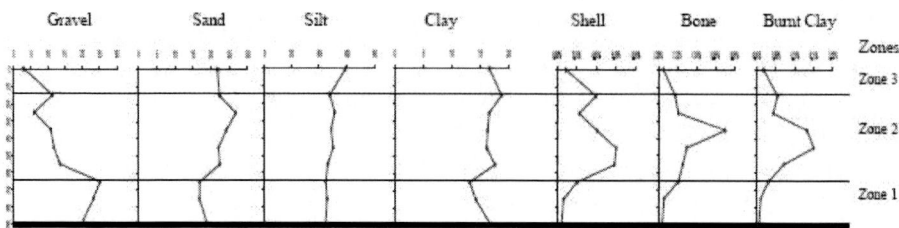

Figure 6.6. Particle size and microartifacts for GP5.

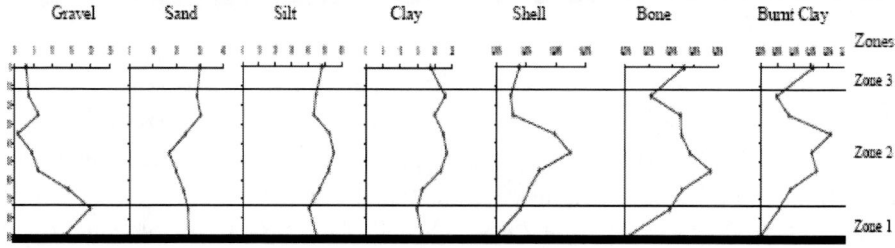

Figure 6.7. Particle size and microartifacts for GP4.

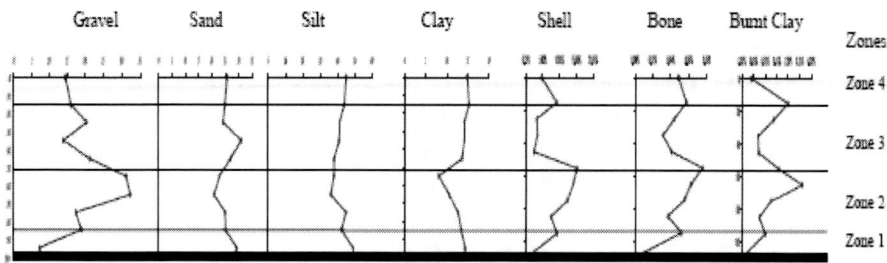

Figure 6.8. Particle size and microartifacts for GP3.

Thus, the geoarchaeological analysis of the deposit covering the Hellenistic theatre at Gitana indicates that:

(a) The abundance of clay particles may be compatible with slow rates of sediment accumulation while the presence of microartifacts denotes its cultural origin from eroded cultural material from the abandoned Hellenistic city, expanded on top of it.

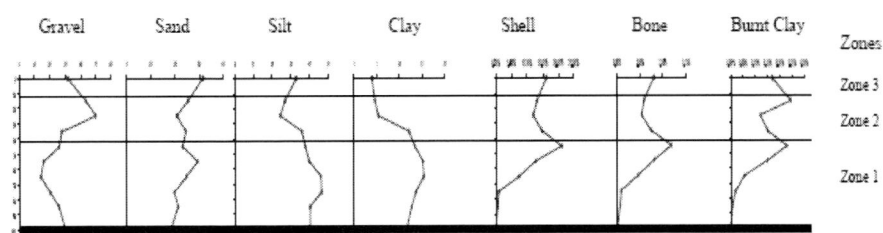

Figure 6.9. Particle size and microartifacts for GP2.

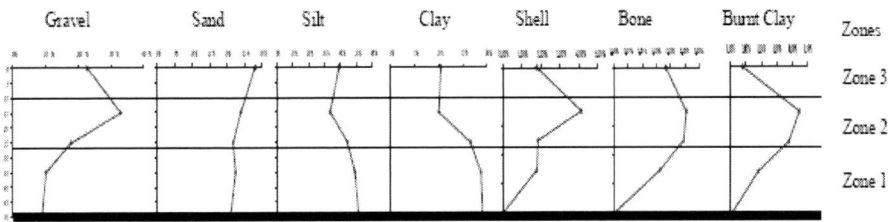

Figure 6.10. Particle size and microartifacts for GP1.

(b) Although is impossible to assess whether microartifacts ended up in the deposit as micro-materials or were generated after deposition, since the identified microartifact types (i.e., shell, bone, burnt clay) come from fragile or size unstable materials, their irregular density trends might have been produced post-depositionally from the effects of different formation processes (i.e., erosion, in situ weathering and translocation of smaller cultural sedimentary particles) that might have affected the larger, macro-artefacts (Sherwood et al., 1995) present in the deposit. In this case, since anthropogenic activity was absent from the site after abandonment, their variability in such fine sediment most likely depicts the intensity of these types of formation processes and possibly the time span capable of producing variable microartifact concentrations.

(c) The peaks of gravel at some points in the sequence are possibly indicative of temporary acceleration of sediment input (mostly gravel sized) and might be related with episodes of collapse of the superimposed city's walls and/or other structures.

(d) The presence of microartifacts in the 'top-soil' layer implies its cultural origin from eroded cultural material from the abandoned Hellenistic city while their variability may be, again, attributed to, for example, *in situ* weathering of these cultural materials.

It seems, therefore, that the formation of the colluvial deposit, covering the abandoned Hellenistic theatre, was a relatively slow process with sporadic acceleration in sedimentary (and arguably artefactual) input. However, it might have been highly shaped by the effects of different natural formation processes.

6.4. CONCLUSIONS

The results from particle size analysis and microartifacts of the colluvial deposit indicate that is composed primarily of silt and clay particles, abundance of sands with additional coarser (gravel-sized) material and cultural sedimentary particles. The erosion of the abandoned Hellenistic city's cultural sediments triggered its formation which was generally a slow process, but the sedimentary and microartifactual characteristics might have been, to a large extend, formed by the effects of post-depositional natural processes.

REFERENCES

ASTM (American Society for Testing Materials) (1961). Tentative method for grain-size analysis in soils. In *The 1961 Book of ASTM Standards*, pt.4, pp.1272-1283.

Birkeland, P.W. (1999). *Soils and Geomorphology*. Oxford: Oxford University Press.

Davidson, D.A. (1973) Particle Size and Phoshate Analysis. Evidence for the Evolution of a Tell. *Archaeometry* 15, pp.143-152.

Folk, R.L. (1980). *Petrology of Sedimentary Rocks*. Austin: Hemphil.

Sherwood, S.C., Simek, J., and Polhemus, R. (1995). Artifact Size and Spatial Process: macro and microartifacts in a Mississippian House. *Geoarchaeology* 10 (6), pp. 429-455.

In: On Site Geoarchaeology on a Neolithic Tell Site ... ISBN 978-1-60741-366-0
Editor: D. Kontogiorgos © 2009 Nova Science Publishers, Inc.

Chapter 7

A COMMENT ON THE VERTICAL MOVEMENT OF MICROARTIFACTS IN AN ANTHROPOGENIC AND IN A NON ANTHROPOGENIC ENVIRONMENT AND THE EFFECTS OF FORMATION PROCESSES

Dimitris Kontogiorgos
Department of Archaeology, University of Sheffield, Northgate House, Sheffield, U.K.

Since Schiffer's (1972) original recognition of the importance of studying and understanding the formation processes of the archaeological record, many authors have pointed out their critical importance (e.g. Goldberg *et al.,* 1993; McGuire, 1995; Reid, 1995; Tani, 1995). Moreover, it is now widely accepted that variability is introduced into the archaeological record through cultural and non-cultural formation processes which distort systemic patterns as well as creating their own patterns (Schiffer, 1987). The unit of analysis appropriate for identifying formation processes is, according to Schiffer (1987) the deposit, but "viewing the deposit as a single discrete depositional event or process has its problems, as a single depositional process can give rise to materials in different deposits, and conversely, a single deposit can contain the products of many different depositional processes" (Schiffer 1987:266).

Despite the recognised importance of cultural and natural processes in the formation of the archaeological record, studies addressing the interpretative potential of microartifacts remain relatively limited, although microartifacts, due to their abundance and incorporation in an archaeological deposit constitute a significant part of the cultural particles present and may provide information on the cultural and natural formation processes occurring in a deposit (e.g., Hassan, 1978; Fladmark, 1982; Vance, 1986; Rosen, 1986; 1989; Dunnell and Stein, 1989; Sherwood, 2001). Dunnell and Stein (1989) outline some of the important characteristics of microartifacts that compel their consideration as archaeological data of the first order. They note, that information content may be different for microartifacts than for larger artifacts and they may be most informative about different things (e.g., particle transport and site formation processes). Equally important, processes that generate microscopic artifacts vary depending on material and context (Dunnell and Stein, 1989: 34-36). These last two issues, differing information content and differing formation processes within the microscale are important reasons for undertaking microartifact analysis (c.f., Dunnell and Stein, 1989).

Then again, attempting to define cultural and natural formation processes in a site focusing, for example, either in their variability or in the proportional correlation among microartifact classes may be misleading because their archaeological significance rests upon understanding the interaction among, the almost, numerous variables within a sequence which would determine their transport potential. The examples offered in the preceding Chapters 2, 5, and 6 of this volume demonstrate variability in microartifacts in deposits more likely to be sensitive in cultural formation processes (as in the case of the Neolithic ditch in Chapter 2 or the EN pit 24 in Chapter 5) but also in deposits formed simply by the effects of natural processes (as in the case microartifacts' variability in the colluvial deposit at the Hellenistic Theatre-Chapter 6). Despite this distance in time, differences in the types of these archaeological contexts, and the different formation processes that have affected them, the cultural micro-material classes display significant variability. This variability cannot simply attributed to the varying friability of the different types of microartifacts (i.e., burnt clay breaks down easier than shell/bone).

The three preceding examples (i.e., Neolithic ditch on the tell, EN pit 24 at the extended settlement, and the sequences from the colluvial deposit covering the Hellenistic Theatre) indicate that similar types of microartifacts within different archaeological contexts exhibit significant vertical mobility as a result of different types of formation processes (cultural and/or natural). This ability for vertical mobility may be used to demonstrate the intensity of these different formation processes; for example, the vertical movement and variability of microartifacts in, the fine textured, colluvial deposit of the Hellenistic Theatre (see Chapter 6) was attributed to the intensity of different natural processes or the variable values for microartifacts in the sequences from the Neolithic ditch (see Chapter 2) or the EN pit 24 (see Chapter 5) were assumed to imply, for the most part, variability in cultural formation processes.

In any case, stronger interpretation can only be achieved by strong microartifact pattern recognition (Kontogiorgos et al, 2007; Kontogiorgos, 2008) especially in cases of archaeological deposits sensitive to cultural formation processes.

REFERENCES

Fladmark, K.R. (1982). Microdebitage analysis: initial considerations. *Journal of Archaeological Science* 9, pp. 205-220.

Hassan, F.A. (1978). Sediments in archaeology: Methods and implications for palaeoenvironmental and cultural analysis. *Journal of Field Archaeology* 5, pp.197-213.

Dunnell, R.C., and Stein, J.K. (1989). Theoretical issues in the interpretation of microartifacts. *Geoarchaeology* 4 (1), pp. 31-42.

Goldberg, P., D. T. Nash and M. D. Petraglia. (1993). *Formation Processes in Archaeological Context*. Madison, Wisconsin: Prehistory Press.

Kontogiorgos, D, Leontitsis, A., and Sangole, A. (2007). Telling a non linear story: The investigation of microartefacts non linear structure. *Journal of Archaeological Science* 34: 1532-1536.

Kontogiorgos, D. (2008). Geoarchaeological and Microartifact Analysis of Archaeological Sediments. A Case study From a Neolithic Tell Site in Greece. Nova Science Publishers, Inc., New York.

McGuire, R. H. (1995). Behavioural Archaeology: Reflections of a Prodigal Son. In J. M. Skibo, W. H.Walker and A. E. Nielsen (eds.) *Expanding Archaeology*, pp. 162-177. Salt Lake City: University of Utah Press.

Reid, J. J. (1995). Four Strategies after Twenty Years: A Return to Basics. In J. M. Skibo, W. H. Walker and A. E. Nielsen (eds.) *Expanding Archaeology*, pp. 15-21. Salt Lake City: University of Utah Press.

Rosen, A.M. (1986). *Cities of Clay: The Geoarchaeology of Tells*. University of Chicago Press, Chicago.

Rosen, A.M. (1989). Ancient Town and City Sites: A View from the Microscope. *American Antiquity* 54, pp. 564-578.

Schiffer, M.B. (1972). Archaeological Context and Systemic Context. *American Antiquity* 37, pp.156-165.

Schiffer, M.B. (1987). *Formation Processes of the Archaeological Record*. University of New Mexico Press, Albuquerque.

Sherwood, S.C. (2001). Microartifacts. In Goldberg, P., Holliday, V.T., and Ferring, R. (eds.), *Earth Sciences and Archaeology*, pp. 327-351, Kluwer Academic/ Plenum Publishers, New York, 2001.

Tani, M. (1995). Beyond the Identification of Formation Processes: Behavioural Inference Based on Traces Left by Cultural Formation Processes. *Journal of Archaeological Method and Theory* 2(3), pp. 231-252.

Vance, E.D. (1987). Microdebitage and archaeological activity analysis. *Archaeology* 40, pp. 58-59.

In: On Site Geoarchaeology on a Neolithic Tell Site ... ISBN 978-1-60741-366-0
Editor: D. Kontogiorgos © 2009 Nova Science Publishers, Inc.

Chapter 8

CONCLUDING REMARKS

Dimitris Kontogiorgos
Department of Archaeology, University of Sheffield, Northgate House, Sheffield, U.K.

The study of Neolithic tell sites in Greece has been benefited by the excellent preservation of architectural features and house remains. Archaeologists studied the spatial dimensions of Neolithic architecture as a basis for the interpretation of the size and possibly the structure of the prehistoric society, social organisation of residential units, and the symbolism and meaning of space (e.g., Kotsakis, 1999). The Neolithic house and household was a central concern for the early Greek farmers and are used in the archaeological analysis as the symbols of the process of neolithization and a settled way of life (e.g., Halstead, 1999).

Under this respect, when architectural preservation is poor, as at the Neolithic tell at Paliambela (e.g., Kotsakis and Halstead, 2004) assessing the processes that shaped the site's form is a difficult undertaken. The analysis of archaeological sediments, building materials and natural sediments presented in Chapters 1 and 2 has shown that the formation of the tell at Paliambela could be largely attributed to these types of cultural products derived from the exploitation of the available natural sources around the site. This type of cultural transformation of the raw materials is enclosed into the building materials, and arguably on the site's sediments. In line with previous discussions on the constraints and possibilities inherent in the mechanical properties of the raw materials utilized by humans that result into the transformation of raw materials into cultural products (e.g., Hughes,

1979; Ingold, 1988, 1990), the analysis of building materials from the tell site at Paliambela could not be more than reflective of this cultural manipulation of the available sedimentary natural sources and their eventual transformation to social products. Viewing the formation of the Neolithic tell at Paliambela under this theoretical perspective, then the effects of cultural formation processes could be upgraded to a major acting force behind its formation.

The thorniness of identifying cultural formation processes from the study of archaeological sediments becomes even harder for the flat-lying, "non-tell" type of settlement that has been recognised as dominant in the Neolithic of Northern Greece. Flat settlements feature expansion by horizontal house replacement and are often considered to provide much less reliable evidence on settlement size and organisation (e.g., Tringham and Krstic, 1990). These type of settlements do not provide good architectural preservation and the interpretations made, as regards to their formation, intrasite organisation, subsistence, and social structure are based, to a large extent, on the allocation of the preserved features (i.e., ditches and subsidiary pits) and the analysis of their contents (e.g., Pappa and Besios, 1999). The analysis of the fill of a large EN pit in the newly excavated flat/extended Early Neolithic site at Korinos (Northern Pieria region, Northern Greece), offered in Chapter 5, has succeeded in identifying cultural formation processes as responsible for its formation. Although, the attempt to further explore the cultural use of this particular archaeological feature and possibly link it to a pit-house, provided unclear results, however, it stressed the necessity for an in depth investigation of the formation processes occurring in these types of archaeological contexts detected at extended Neolithic settlements.

The ability to unfold the information encoded into archaeological sediments through the application of geoarchaeological techniques is the key to understanding and interpreting the formation processes in deposits of cultural material. Without underestimating the effects of natural processes or rather 'naively' expecting cultural factors to account for all the extant variability in an archaeological site, it seems that drawing logical connections between geoarchaeological data and past human activities upgrades and enhances cultural interference upon natural factors in a site's formation. The application of computational techniques, as the ones offered in Chapters 3 and 4, seems to be more than a necessity in modern geoarchaeological studies. The study of microartifacts, those cultural particles included into archaeological sediments, although by no means conclusive, can be utilised to identify forms of behaviour enacted within a site, when strong pattern recognition has been achieved (e.g., Kontogiorgos, 2008) while it could be also useful for observing variability in natural formation processes, as in the case of microartifacts detected in the colluvial deposit covering the Hellenistic theatre at Gitana (Epirus, NW Greece) (Chapter 6)

New ways of describing variation in archaeological assemblages could only be effective if we could connect them with past human behaviour in a non static physical environment. The methods used and presented in the studies of this volume to quantify the characteristics of archaeological sediments and the theoretical orientation, have highlighted some of the information encoded within those sediments. It is through such methods and theories as utilized here that will further the archaeological community in our ability to interpret the archaeological record and to objectively assess the accuracy of our interpretations.

REFERENCES

Halstead, P. (1999). Neighbours from Hell? The Household in Neolithic Greece. In *Neolithic Society in Greece*, edited by Halstead, P., pp. 77-95. Sheffield Studies in Aegean Archaeology 2, Sheffield Academic Press.

Hughes, T. (1979). The electrification of America: the system builders. *Technology and Culture* 20(1), pp. 124-162.

Ingold, T. (1988). Tools, minds and machines: an excursion in the philosophy of technology. *Techniques ae Culture* 12, pp. 151-176.

Ingold, T. (1990). Society, nature and the concept of technology. *Archaeological Review from Cambridge* 9(1), pp. 5-18.

Kontogiorgos, D. (2008). Geoarchaeological and Microartifact analysis of Archaeological Sediments. A Case Study from a Neolithic Tell Site in Greece. Nova Science Publishers, Inc. New York.

Kotsakis, K. (1999). What Tells Can Tell: Social Space and Settlement in the Greek Neolithic. In *Neolithic Society in Greece*, edited by Halstead, P., pp. 66-76. Sheffield Studies in Aegean Archaeology 2, Sheffield Academic Press.

Kotsakis, K., and Halstead, P. (2004). Anaskafi sta Neolithika Paliambela Kolindrou. In *Arhaeologiko Ergo sti Makedonia kai sti Thraki* 16, pp. 407-415 (2002). Thessaloniki: Ministry of Culture.

Pappa, M., and Besios, M. (1999). The Makriyalos project: Rescue Excavations at the Neolithic Site of Makriyalos, Pieria, Northern Greece. In *Neolithic Society in Greece*, ed. Halstead, P., Sheffield Academic Press, 1999, Sheffield.

Tringham, R., and Krstic, D. (1990). *Selevac: A Neolithic Village in Yugoslavia* (Monumenta Archaeologica 15). Los Angeles: Institute of Archaeology Press, UCLA.

INDEX

A

accuracy, 51, 87
adjustment, 51
age, 68
agricultural, 66
algorithm, 28, 29, 45, 46, 48, 49
alluvial, 2, 3, 5, 12
anthropogenic, 79
application, 3, 28, 46, 49, 55, 56, 86
Aristotle, vii
ASTM, 3, 16, 21, 24, 63, 67, 73, 80
Athens, vii, 69, 71
autocorrelation, 49
availability, 4
awareness, 67

B

Balkans, 1
behavior, 48
birds, 48
bounds, 47
Bulgaria, 1
buttons, 53

C

classes, 28, 46, 47, 82
clay, 3, 9, 11, 12, 21, 22, 23, 29, 30, 60, 62, 63, 65, 74, 75, 76, 78, 79, 80, 82
coal, 31, 32, 34, 36, 38, 40, 42
colors, 53
communication, 48, 49
community, 87
competition, 58
components, 19, 59
composition, 3, 12, 15
concentration, 1, 22
confidence, 28, 37, 44, 45, 64
confidence interval, 28, 37, 44, 45, 64
confidence intervals, 28, 37, 44, 45, 64
Congress, 56
constraints, 1, 85
construction, 2, 58
continuity, 19, 58
conversion, 11
correlation, 82
covering, 29, 79, 80, 82, 87
cultural factors, 86
cultural transformation, 85
culture, 59
cumulative frequency, 11
cumulative percentage, 11

D

data set, 4
definition, 58
density, 3, 22, 23, 63, 64, 66, 79
deposition, 19, 24, 65, 66, 79
deposits, 2, 4, 5, 9, 11, 12, 15, 16, 19, 24, 28, 29, 48, 55, 58, 59, 67, 81, 82, 83, 86
determinism, 28, 48
discontinuity, 19, 58
discrimination, 11, 13
distilled water, 3
distribution, 4, 11, 12, 17, 44
drying, 4

E

earth, 2
Earth Science, 16, 84
e-mail, 27, 47
encoding, 30
encouragement, vii
erosion, 79, 80
estimating, 27, 45, 47
Europe, 1, 2, 16, 18, 60, 67, 68, 72
evolution, 29
exercise, 30
exploitation, 85
exposure, 66
extraction, 59

F

family, 57
farmers, 85
flight, 49
focusing, 82
food, 48
forecasting, 45
freedom, 44
frequency distribution, 17
funding, vii

G

generation, 55, 77, 78
genetic algorithms, 46
glass, 29
grain, 16, 17, 24, 67, 80
graph, 17, 30
Greece, 1, 2, 16, 17, 19, 24, 27, 29, 47, 57, 58, 59, 60, 67, 68, 69, 71, 72, 83, 85, 86, 87

H

habitation, 59, 66
heart, 48
heuristic, 45
higher quality, 4
Holocene, 3, 5, 12
household, 58, 85
human, 11, 12, 20, 59, 66, 86, 87
human activity, 11, 12
humans, 85
hybrid, 45, 46
hyperbolic, 4

I

identification, 47, 58
in situ, 79, 80
indication, 78
indicators, 11
inferences, 58
interaction, 82
interface, 49
interference, 86
iteration, 49

J

justification, 4

Index

K

Kolmogorov, 4, 17

L

laboratory method, 20, 59, 71
Late Quaternary, 68
law, 4, 11
likelihood, 11, 44
linear, 53, 83
lying, 86

M

Macedonia, 29, 68
machines, 87
manipulation, 86
matrix, 30, 49, 53
measurement, 4, 16, 30
mechanical properties, 85
microscope, 29, 30
Microsoft, 48, 49, 50, 51, 53
misleading, 82
mobility, 69, 82
models, 4
movement, 48, 49, 83
mutation, 28, 29, 30, 46

N

natural, 2, 3, 5, 12, 13, 16, 29, 30, 31, 37, 44, 59, 80, 82, 85, 86
next generation, 30
Nielsen, 4, 16, 68, 83
noise, 11
normal, 1, 4
normal conditions, 1
normal distribution, 4

O

optical, 29, 30
optimization, 48, 51, 53, 54, 55, 56
organic, 2, 29
organic matter, 29
orientation, 87

P

parameter, 53
particle swarm optimization, 55, 56
particles, 3, 11, 12, 16, 21, 23, 28, 29, 30, 31, 37, 44, 48, 49, 51, 53, 63, 73, 76, 77, 79, 80, 82, 86
pattern recognition, 83, 86
patterning, 59, 67
petrology, 17, 18, 24, 67, 80
philosophy, 87
physical environment, 87
planning, 55
poor, 51, 85
population, 28, 29, 30
population size, 28
pragmatic, 4
probability, 17, 28, 29, 30, 44
program, 48, 51

R

rainfall, 66
random, 29, 48, 49
random numbers, 49
range, 11
raw material, 85
raw materials, 85
reading, 4
recognition, 81, 83, 86
reflection, 59
representative samples, 27, 47
residential, 85
Romania, 1, 68

S

sample, 3, 4, 5, 6, 7, 8, 9, 10, 12, 29, 30, 44, 45, 47, 49, 51, 53, 54, 55, 61
sampling, 20, 21, 59, 61
sand, 3, 4, 9, 11, 12, 17, 21, 22, 29, 62, 63, 74, 75, 76
search, 29, 48
searching, 48
sediment, 3, 21, 29, 63, 65, 66, 72, 76, 77, 78, 79
sedimentation, 3, 11, 20, 23, 65, 77, 78
sediments, 2, 3, 4, 12, 13, 16, 17, 20, 24, 29, 45, 58, 66, 71, 76, 77, 80, 85, 86, 87
seeds, 59
Serbia, 1, 67
settlements, 1, 58, 86
shape, 11
sharing, 68
shell, 23, 31, 32, 34, 36, 37, 38, 40, 42, 50, 63, 64
shelter, 58
sites, 1, 28, 58, 85
skewness, 11
social structure, 86
sodium, 3, 29
software, 48, 55
soil, 61, 66, 73, 78, 80
soils, 16, 24, 67, 80
sorting, 27, 47
spatial, 19, 58, 85
spheres, 11
sporadic, 80
stability, 1
stages, 66
standard deviation, 11
standard error, 48
standards, 16, 24, 67, 80
statistical analysis, 17
statistics, 11
stochastic, 28, 46
stock, 45
storage, 60, 67
subsistence, 86
swarm, 55
symbols, 85

T

temperature, 3, 4
temporal, 19, 58
Thessaloniki, 46, 87
three-dimensional, 58
time consuming, 49
transformation, 4, 85
translocation, 79
transport, 82

V

values, 22, 24, 28, 44, 45, 53, 65, 77, 78, 83
variability, 20, 22, 24, 57, 58, 59, 61, 66, 77, 79, 80, 81, 82, 83, 86
variables, 82
variation, 19, 28, 45, 59, 65, 87
vector, 30, 49
velocity, 49
village, 60
viscosity, 4
visible, 51

W

water, 3
weathering, 79, 80
workers, 11

Z

Zone 1, 22, 65, 66, 76, 77, 78
Zone 2, 22, 65, 66, 77, 78
Zone 3, 22, 65, 66, 77, 78